Capitalism and Competition: Oil Industry Divestiture and The Public Interest

Proceedings of the
Johns Hopkins University
Conference on Divestiture

Washington, D.C.
May 27, 1976

Sponsored by the
Center for Metropolitan
Planning and Research
and School of Advanced
International Studies
of Johns Hopkins
University

Editors:
George A. Reigeluth
Douglas Thompson

54176

International Standard Book Number 0-917862-01-5
Library of Congress Catalog Card Number 76-46906

Contents

Introduction:

On May 27, 1976, The Johns Hopkins University sponsored a one-day conference at its School of Advanced International Studies in Washington, D.C. on "Capitalism and Competition: Oil Industry Divestiture and the Public Interest." The University has traditionally taken an active part in encouraging informed debate on a wide variety of public policy issues. Recent Congressional interest in divestiture, and the potentially significant impact divestiture could have on industrial structure and the economy in general, led the University to organize this conference. The issues surrounding the concept of divestiture are significant and far-reaching. They range from competition in the industry, the price of energy, and U.S. dependence on foreign oil, to the availability of additional sources of energy and the formation of new capital necessary for future energy exploration and development. Faced with proposals to alter the industrial structure of the country, we must understand the full consequences of such proposals as completely as possible.

Toward this end, the University has decided to publish the proceedings of the conference. The basic purpose of the conference was to provide an arena in which both sides of the many divestiture-related subjects could be discussed and debated. The University hoped to offer a balanced presentation of the divestiture question, to provide time for both antagonists and protagonists to present their views and challenge each other's conclusions.

The intent was not to formulate conclusions with respect to divestiture, but to spell out as fully as possible the consequences of future decisions.

The conference proceedings opened with an introduction placing corporate power and structure in an historical perspective by Professor Alfred H. Chandler, Jr., followed by two views of vertical and horizontal divestiture. The luncheon address, given by Gerald L. Parksy, presented the findings of a recent U.S. Department of the Treasury study on divestiture. The next part of the proceedings covered the afternoon session of the conference. First, two economists discussed and analyzed the effects of divestiture and second, the international perspective was discussed by two participants with reference to the international oil market. The proceedings of the conference closed with a discussion of divestiture and public policy by John C. Sawhill, President of New York University. In addition, selected comments from the panelists are also included.

We hope that the publication of these proceedings will stimulate further debate and discussion. The conference and publication were made possible through support from the American Petroleum Institute and with organizational assistance from the University's Center for Metropolitan Planning and Research.

Steven Muller
President
Johns Hopkins University

Participants In The Conference

Speakers

Steven Muller
President, The Johns Hopkins University

Robert E. Osgood
Dean, The Johns Hopkins School of Advanced International Studies

Alfred H. Chandler, Jr.
Professor, Harvard University

Walter S. Measday
Chief Economist, Subcommittee on Antitrust & Monopoly United States Senate

Annon M. Card
Senior Vice President, Texaco Inc.

Floyd Haskell
Senator, State of Colorado

William T. Slick, Jr.
Senior Vice President, Exxon Company, U.S.A.

Gerald L. Parsky
Assistant Secretary for International Affairs, U.S. Department of Treasury

Edward W. Erickson
Professor, North Carolina State University

James M. Patterson
Professor, Indiana University

Thomas R. Stauffer
Professor, Harvard University

Clement B. Malin
Assistant Administrator for International Energy Affairs, Federal Energy Administration

John C. Sawhill
President, New York University

Panelists

Robert E. Shepherd
  Director, Office of Energy
  Policy & Programs
  U.S. Department of Commerce

William Smith
  Financial and Business
  Correspondent,
  The New York Times

Lee Richardson
  President, Consumer Federation
  of America

Albert J. Anton, Jr.
  Partner, Carl H. Pforzheimer & Co.

John Wallach
  State Department Correspondent,
  Hearst Publications

Stephen H. Goodman
  Vice President, Policy Analysis
  Export-Import Bank

William A. Johnson
  Professor, George Washington
  University

Theodore H. Moran
  Professor, The Johns Hopkins
  University School of Advanced
  International Studies

Moderator: Francis X. Murray
  Associate Director for National
  Energy Programs Center for
  Strategies and International Studies

1 DIVESTITURE IN HISTORICAL PERSPECTIVE

Alfred H. Chandler, Jr.
Professor
Harvard University

I will confess to being a professional historian and as a historian I am rather short on current information but somewhat longer on perspective. What I would like to do this morning is review briefly the history of the oil industry, particularly its corporate organization, and then relate the experience of the oil industry to that of other American industries. The oil industry is the oldest and one of the most representative of the modern mass producing, mass marketing industries.

The oil industry was one of the very first to perfect modern, large batch and then continuous process high-volume technology; and it was one of the very first to sell its products in national and global markets. The major market for oil in the early years of the industry was kerosene for lighting and cooking, and its major market was abroad.

Most American-made kerosene was sold in Europe. Oil sales to Europe were 80 percent more than U.S. domestic consumption. There the economy was already becoming urban and industrial. The United States was still rural. People went to bed with the chickens and they got up with the chickens. They were still amply supplied with wood. So there was less need for kerosene in the United States.

The technology of production by 1869, ten years after oil was discovered at Titusville, was already close to continuous process. From the time it entered the refinery until it came out as kerosene, petroleum was untouched by human hands. Labor was used only for packaging. Further economies were achieved in these early years through the use of superheated steam and of cracking methods. Steam was also used to move oil through the refinery.

These innovations greatly increased the volume and speed of

throughput. The faster the petroleum went through the refinery, the lower the unit cost and the greater the competitive advantage. One does not have to be an economic historian to know who had the fastest refinery in the west in 1869. So I will begin, as must any historical account of the oil industry, with John D. Rockefeller and the Standard Oil Company. Mine differs from the usual accounts by emphasizing markets and technology rather than robber barons and industrial statesmen.

Rockefeller used the high-volume output and low unit costs of the fastest refinery in the west to obtain a monopoly. He did it in the following way. He sent his man Flager to Devereaux, the General Manager of the Lake Shore Railroad, and offered to provide that road with sixty carloads of petroleum a day, every day. Standard Oil would supply the tanks in which the kerosene was stored. It would do this if Devereaux gave the company a special rate which was half the official listed one. Devereaux was delighted. All he had to do was to supply a locomotive and sixty flat cars a day. He made more money with the new rate under the new requirements than he had when oil was carried in a more haphazard and expensive manner at the old rate. Rockefeller and his associates then took their rate agreement to the leading refiners in Cleveland where the Standard Oil refinery was located, and suggested that they join together to take advantage of the rate he had acquired. After an easy conquest of Cleveland, the Standard Oil used the same strategy to make allies of the other leading refineries in the nation. Thus by controlling transportation, the Standard Oil Company and its associates soon had a monopoly, that is 90% of the refining production in the United States. The cartel of 25 to 30 companies controlled the industry through their control of transportation. But this method of control did not last long. Technology brought the change.

The producers, to break Standard Oil's hold on transportation, completed, in May 1879, the very first long-distance pipeline. At that time all production of crude oil was concentrated in western Pennsylvania. The producers formed the Tidewater Oil Company to build a pipeline across the mountains to a railroad not controlled by the Standard Oil group. Rockefeller and his associates fought the pipeline all the way in the classic tradition. They purchased land ahead of the line, brought injunctions against its construction and even tore up a line at night after it had been laid down the day before by the competitors' men. Nevertheless, the Tidewater Company did complete the pipeline to Tidewater. Soon it had its own refinery and was marketing in Europe.

The response of the Standard Oil group was quick. The owner-managers of the allied companies quickly realized the great reduction in cost permitted by the pipeline. If unit costs were reduced by increasing the speed of throughput in a refinery, they would be more greatly reduced if the flow could be carefully scheduled from the oil wells to the refinery. Not only were transportation costs lower but the pipelines themselves provided storage areas.

To realize the economic advantages permitted by the new long-distance pipeline, however, required a reorganization of the whole refining industry. Since crude oil would be piped but the refined product could not yet be moved in that way, it was logical to concentrate refining at major shipping points. Full and steady flow of supplies meant that much larger refineries could be constructed.

A cartel, however, could not make the decisions to build and operate pipelines or new refineries. A cartel was merely an association of many independent firms. A new legal form had to be devised to permit the Standard Oil associates to make the investment decisions involved in constructing the pipelines and then in shutting down old refineries and building new ones. The new form was the Trust. The Standard Oil Trust was a legal device by which a management board — a board of trustees — could make these investment decisions, carry them out, and then manage the properties.

Rockefeller and his associates moved last. Late in 1881 the trust agreement was signed. The refining facilities were then consolidated. By 1885 the number of major refineries were reduced from 53 to 22. Two-fifths of the world's production of oil came to be carried out in three refineries, one at Bayonne, New Jersey, another at Philadelphia, and a third in Cleveland. These were major shipping points — two for Europe and one for the American west. This concentration of production meant that costs dropped from one-and-one-half cents a barrel to a half-a-cent a barrel.

These cost reductions reflect the nature of modern continuous process technology. Imagine putting two-fifths of the world's shoe production into three factories. The diseconomies of scale would have been enormous. On the other hand the economies of such concentration were striking in oil refining.

Its much greater volume of production, refined at much lower unit costs, pushed the Standard Oil Trust forward into marketing. Its board of trustees soon set up their own distribution companies or purchased existing large wholesalers in the United States and in Europe. By the second half of the 1880s they had created a national and global marketing and distributing organization.

Then at the very end of the decade, when the Pennsylvania oil fields began to decline and new fields came in at Lima, Indiana, the Standard Oil Trust became concerned, for the first time, about the availability of supplies. The trustees then decided to move into the production of crude oil. By 1892 Standard Oil was producing one-fourth of the nation's crude oil.

In the meantime, as the Pennsylvania fields began to dry up, the producers were finally able to consolidate and obtain some control over output. A number of them organized the Pure Oil Company, then built another pipeline across the mountains, set up a refinery on the coast, and began to market at home and abroad. Thus by the 1890s the structure of the oil industry had acquired what would become a standard pattern in a number of American industries. One giant integrated firm competed with two smaller independents which were also integrated.

The situation, however, was not yet stable. This time changing markets and supplies rather than technology generated change. In the 1890s there was little cheer in the headquarters of Standard Oil at 26 Broadway, nor in those of Tidewater and Pure Oil. Kerosene was being challenged by a new type of lighting. In 1882 Thomas Edison had established the first central electric power station at Pearl Street in New York City. By 1892 two giant enterprises — General Electric and Westinghouse — were turning out vast quantities of electrical producing equipment. By the end of the century the market for kerosene was beginning to decline.

At that same moment, however, a totally fortuitous circumstance saved the petroleum industry. The automobile had become more than a rich man's toy. In 1900 Ransom Olds began to manufacture cars for the commercial market. By 1908 Henry Ford was beginning to mass produce the Model-T Ford. Almost overnight a vast new market had appeared.

At the same time the supply situation dramatically changed. Great new fields were opened up in Texas and in California. As a result the first decade of the 20th Century was the most dynamic in the history of the oil industry. New giant companies appeared to exploit the new production fields and to supply the new gasoline market.

The accompanying Table I illustrates this point. Well before the break-up of Standard Oil in 1911 for violating the Sherman Antitrust Act, several large integrated competitors were already in operation. Six of these companies had been established after 1900. They were, by 1911, already among the largest business enterprises in the United States. The Texas Company (it became Texaco) was already the 24th largest and Gulf already the 26th largest industrial company in the country. And the companies coming out of the California and Texas fields, like the early ones in Pennsylvania, quickly integrated forward into refining and then into marketing.

Then in 1911 came the first great divestiture. The Standard Oil Company of New Jersey, the successor to the Standard Oil Trust Company, was dismembered. The Standard Oil Company, like the Trust, was essentially a holding company which operated its activities through functional subsidiaries. There were refining subsidiaries, marketing subsidiaries, crude oil subsidiaries, and transportation subsidiaries. Only three or four Standard Oil firms were even partially integrated. Thus when divestiture came, it was in a sense a vertical divestiture in that the different functions in the industry became carried out by non-integrated companies. Nevertheless, the refining firms quickly moved forward into marketing and backwards into production and the marketing companies moved backwards into refining. Standard Oil of New York, for example, operated in 1911 a great marketing empire. It quickly built its own refineries, although it did not go into the production of crude oil until after 1917.

By 1917, therefore, the modern structure of the industry had begun to appear. It would become fully established after World War I. From the 1920s on 20 or so large integrated enterprises carried on the giant share of the oil industry's activities. Many of these firms grew by absorbing smaller ones. The right-hand column of the accompanying table suggests this process. Only one of the major firms founded between 1911 and 1917 remained an independent. That one was Sinclair. The rest merged into firms in the first and second columns of the table. The names in brackets after each company indicate the enterprise into which it was absorbed. The names on this table are still today those of the industry's major companies. (Two

— Phillips and Continental — one an independent and the other a former Standard Oil Company — were operating in 1917 but are not listed in Table I as they did not yet have assets of $20 million.)

In the 1920s these firms were still primarily domestic enterprises. They produced for regional markets in the United States and obtained most of their crude oil from American sources. The two exceptions were Standard Oil of New Jersey and Standard Oil of New York (Socony), both of which had large overseas marketing organizations created in earlier years.

The 1930s brought two developments. First, the great depression caused oil companies to begin to diversify out of the gasoline market into fuel oil, household heating, diesel oil and even aviation gasoline. At the same time a few firms began to reach overseas. The most active was Standard Oil of California which started to market in Asia in a joint venture with Texaco. It also entered into the new Middle East crude oil fields. Socony, after its merger with Vacuum Oil in 1931, joined Jersey Standard in a joint venture to produce and market oil east of Suez. Gulf began to look to the Middle East for crude supplies, but continued to market its products largely in the United States.

The years after World War II were also marked by two major developments. Unlike the 1930s, the market expanded at an unprecedented rate. This meant a vast increase in refining and a major expansion of marketing abroad. The great growth of national and global markets further enlarged the demand for crude oil and brought more companies into overseas production. The five firms that had already gone abroad before the war now had a substantial advantage. Thus Jersey Standard, Socony, Mobil (the successor to Socony and Socony-Vacuum), Texaco and Standard of California became more dominant in the industry than they had before the war.

Secondly, the oil companies after the war began to diversify into new product markets. Before 1941 they had paid little attention to petrochemicals. They were uncertain of the technology, and the market for petrochemicals differed from that for gasoline. Companies like Standard of Indiana had chemical firms build plants next to their refineries to further process petrochemicals.

During the war the oil companies learned much about the technology of making petrochemicals. The synthetic rubber program in particular brought the oil companies into chemistry. At the same time the market of petrochemicals grew much larger with the development of plastics and other new products. Therefore, high-volume production and marketing on a national and global scale appeared for these chemicals for the the first time. By the 1950s most oil companies had moved into petrochemicals and had set up separate autonomous operating divisions to market, produce and develop petrochemicals. In the post-war years, too, the oil companies began to take up the production and distribution of other sources of energy besides petroleum.

This then is a very brief history of the corporate structure of the oil industry. Its history is typical of many American industries. Other industries started with a horizontal combination which became a vertically integrated enterprise. Often there was one giant firm and a small number of integrated independents. Then as time passed and markets grew, the number of firms operating in the oligopoly increased. And like the oil enterprises, those in other industries began during and after World War II to diversify into new product markets and to expand their activities overseas. This pattern occurred in primary metals, chemicals, rubber, food, automobiles, electrical machinery and other machinery.

The second of the attached tables suggests this pattern. This table indicates the location of all American firms (278 of them) producing goods which had assets of $20 million or more in 1917. Such enterprises cluster in manufacturing. Of the 278 there are only 5 in agriculture, none are in construction, 30 are in mining and 7 in crude oil. All the rest are manufacturing or processing enterprises. And nearly all operated in industries which were dominated by a small number of large vertically integrated manufacturing or processing firms.

These oligopolistic industries have been, since modern industrialization began, the most vital to the continuing growth and health of the American economy. Such industries have never been competitive in the traditional sense. On the other hand, the labor-intensive competitive industries such as furniture, lumber, printing and publishing, leather, apparel, and textile, have been less central to the economy's growth and health. And what is true of the United States is true in other advanced industrial countries. In Great Britain, France, Germany and Japan and large firms appeared in very much the same industries. There the dominant firms have also become integrated enterprises. The historical lesson would seem to be that the large vertically integrated enterprise competing oligopolistically is not some sort of unhealthy aberration. Such enterprises have been the standard institutional response to modern technology and modern markets in all advanced market economies.

Now let me say a final word on divestiture. From the start Americans have been distrustful of these giant business enterprises. The concentrated economic power they acquired appeared to violate basic values. Such power was hardly democratic. It restricted the opportunities of enterprising individual entrepreneurs. Since the beginning of this century there has been great public concern and strong political protest against this brand new kind of modern institution.

In the century's first decade the control of the large corporation became the most significant domestic political issue. It was the central concern of the American voters in the Progressive era, the age of Theodore Roosevelt and Woodrow Wilson. A second wave of protest came a generation later in the late 1930s during Franklin Roosevelt's New Deal. Now in the mid-1970s we see a third wave. This underlying concern with concentrated economic power seems to peak once every generation.

The oil industry's experience during these three peaks has been representative. In 1911 came the breakup of Standard Oil, the first great vertical divestiture. In the 1930s the Justice Department brought suit against 22 large oil companies again calling for divestiture. At that time the goal was to remove the pipeline from the integrated companies, as the control of pipeline transportation appeared to be central to the maintenance of their economic dominance. And now, in 1976, Congress is calling once again for divestiture.

I suspect that in today's papers and discussions, we will hear much the same arguments pro and con as those used by our grandfathers in the Progressive era and by our fathers during the New Deal. Indeed, probably many of the same solutions will again be offered. We should keep in mind, however, that there is one difference between ourselves and our fathers and particularly our grandfathers. Our grandfathers had only 20 or 30 years of looking at the operation of a modern industrial economy. We have had the benefit of 80 and 90 years of experience. We really should give some consideration to this experience.

Its most obvious meaning is that, since the beginning of modern mass production and modern mass distribution, the large integrated enterprises have been at the center of economy. Such enterprises have managed the basic processes of production and distribution in the United States since the 1880s. The historical experience also suggests that markets and technology determine whether an industry is concentrated or not. Concentration is not the work of robber barons or industrial statesmen. Nor is it the result of public legislation. Markets and technology and not antitrust laws have determined why the auto, rubber and oil industries have almost always been concentrated and that the furniture, apparel, and the leather industries have almost never been concentrated.

Clearly the passing of laws will not readjust the fundamental structure of a modern industrial economy. Control over these powerful industrial units poses one of the most serious challenges of our times. Except for the discipline of obtaining a continuing steady rate of return investment, the managers of these enterprises are, in fact, not responsible to anyone. Boards of directors have little to say in management decisions, and stockholders none at all.

I, for one, hope that today's discussion will suggest some ways of controlling or at least providing a check on these powerful enterprises. I look for answers more imaginative and more effective than those of divestiture which our grandfathers proposed so long ago. As a historian, however, I am skeptical. We rarely learn from history. Economists, as Dean Osgood has pointed out, do not look to the past. Businessmen are too busy. So too are men in politics. Nevertheless, I will close by reminding you of the words of the sage: "Those who do not understand history are bound to repeat it."

Table 1

Petroleum Companies in 1917 with Assets of over $20 Million

Standard Group	Independents Before 1911	Independents After 1911
2. Standard Oil Co. of N.J.	24. Texas Co.	37. Magnolia Petroleum Co. (no c) [Socony]
14. Standard Oil Co. of N.Y. (no c)* [Socony]	26. Gulf Oil Co.	
	45. Pure Oil Co. [Union Oil-1965]	56. Sinclair Oil & Refining Corp.
34. Standard Oil Co. of Ind.	69. Associated Oil Co. [Tide Water]	64. Pan American Petroleum & Transport Co. [Standard Indiana]
35. Standard Oil Co. of Calif.	71. Union Oil of Calif.	
48. Prairie Oil & Gas Co. (c only) [Sinclair]	124. Tide Water Oil Co.	95. Midwest Refining Co. [Standard Indiana]
61. Ohio Oil Co. (c only)	160. Shell Oil of Calif.	110. Cosden & Co. [Sunray-Mid Continent 1955]
72. Vacuum Oil Co. (no c) [Socony]	229. Sun Co.	151. California Petroleum Corp. (c only) [Texas Co.]
84. Atlantic Refining Co.		
106. Pierce Oil Corp. (liquidated)		162. Texas Pacific Coal & Oil Co. (c only) [Seagrams - 1965]
205. South Penn Oil Co. (c only)		168. Houston Oil Co. of Texas (c only) [Atlantic Refining]
262. Standard Oil Co. (Ohio) (no c)		178. General Petroleum Corp. [Socony]
		261. Producers and Refiners Corp. (no marketing) [Sinclair]
		278. Skelly-Sankey Oil Co. (c only) [Getty Oil - 1967]

---

* *"c" is crude oil. Dates are given for post-World War II mergers. Current names: Socony Vacuum is Mobil Oil; Ohio Oil is Marathon; and South Penn is Pennzoil United.*

Table 2

The Location of the Largest Industrial Enterprises

Of the 278 enterprises involved in the production of goods in the U.S. in 1917 with assets of $20 million or over:

30 in mining   0 in construction
7 in crude oil   236 in manufacturing
5 in agriculture

Of the 236 manufacturing firms:

171 (72.5%) clustered in 6   23 (9.7%) scattered in 7 groups
     two-digit SIC groups

39 in primary metals   7 in textiles
34 in food   5 in lumber
29 in transportation equipment   4 in leather
24 in machinery   3 in printing and publishing
24 in petroleum   3 in apparel
21 in chemicals   1 in instruments
     0 in furniture

The remaining 42 were in continuous-process and large batch four-digit industries within the 7 remaining groups. In the paper group, the large firms were clustered in the production of newsprint and kraft paper; in stone glass, and clay: in cement and plate glass; in rubber: in tires and footwear; in tobacco: in cigarettes; in fabricated metals: in cans; in electrical machinery: in standardized machines; and in miscellaneous: in matches.

2 VERTICAL DIVESTITURE: OPPOSING VIEWS

THE CASE FOR VERTICAL DIVESTITURE

Walter S. Measday
Chief Economist
Subcommittee on
Antitrust & Monopoly
U.S. Senate

Just a year ago this time one of the biggest jokes in Washington was that several Senators and Representatives had put in a variety of bills which would restructure the petroleum industry in one way or another. As recently as late September when Senator Bayh introduced — for himself, Mr. Philip Hart, Mr. Abourezk, Mr. Packwood, and Mr. Tunney — the original version of S. 2387, it was still a subject for high humor in places like the Monocle, the 116 Club, and other bistros in which one is likely to bump into his friends from the oil industry.

Less than three weeks later, the mood changed abruptly. A vertical divestiture amendment to the natural gas deregulation bill in the Senate lost by only five votes, 54-45; a lobbyist remarked to me after the vote — "They'll be drinking double martinis all night at the Petroleum Club in Houston."

On April 1, 1976, the Senate Subcommittee on Antitrust and Monopoly voted to report out favorably an amended version of S. 2387. The full Judiciary Committee has agreed to a vote on the bill on June 15. Suddenly, vertical divestiture of the petroleum industry is a deadly serious matter: It is a deadly serious matter to the industry — and it is equally serious to those people who support divestiture. It is not a cloud which will blow away with the first good breeze.

As reported out of the Subcommittee, S. 2387 envisions a three-way split of vertically-integrated operations. Some 18 major companies — those with domestic production of 100,000 b/d or more, or with domestic refinery input or marketing in excess of some 300,000-plus b/d would be required to separate — worldwide — their crude production from their refining and marketing operations. Within the United States all oil companies would be required to divest their oil pipeline interests.

The oil companies have asked, quite reasonably, "Why us?" They point out that there are other more concentrated industries in the economy. Indeed, they have been more than willing to list these other industries for us.

One answer lies in the enormous importance of the petroleum industry to the U.S. economy. Sales of petroleum refining companies reported by the Federal Trade Commission for 1975 came to more than $130 billion. This can be compared to $74 billion for motor vehicles and equipment, $87 billion for all of the chemical industries, or $66 billion for primary metals industries, both ferrous and non-ferrous. Only the whole broad sweep of industries manufacturing food and kindred products had sales greater than those of the petroleum refining industry. More than this, of course, what happens in the petroleum industry directly affects the functioning of every other industry in the economy and every aspect of our daily lives. If one were forced to choose a single key industry, it would have to be petroleum.

We may grant that concentration ratios at various levels of the petroleum industry are lower than they are in some other industries, but they are still significant. We've heard a great deal, for example, about the 10,000 crude oil producers in the industry. On a gross operator basis in 1974, the four largest companies produced one-third of our output of crude oil and condensate; the eight largest, 54 percent; and the twenty largest, more than 72 percent. I estimate that the 18 companies which appear to come under the provisions of S. 2387 have 75 percent of the nation's production. Moreover, concentration has been increasing rapidly over the past two decades. Concentration at the 20-company level increased from 56 percent in 1955 to 77 percent by 1974. The eight largest producers today have roughly the same share of the market as did the twenty largest in 1955. Further, the future doesn't look too bright for most of those

10,000 crude producers. According to FTC studies based on 1970 data, the eight largest companies control about 65 percent and the twenty largest more than 90 percent of total United States and Canadian proved reserves.

At the refining level, some 154 companies reported refining establishments to the Bureau of the Census for 1972. Concentration ratios have been remarkably stable over the past twenty years, with the four largest consistently holding about a third of the market, the eight largest about 56 percent, and the twenty largest about 84 percent. The stability itself may be surprising, considering the rate of market growth. And there has been very little really new entry into refining in the past quarter century — new companies have come in, but in almost every case they have done so by acquiring existing refiners, with established crude oil sources and outlets for their products.

Nevertheless, we concede — particularly with respect to refining — that concentration is moderately, rather than extremely, high. It is this fact which gives us grounds for believing that restructuring will improve the competitive performance of the industry.

The problem in the oil industry is that what concentration exists is effectively multiplied by other institutional factors which affect the degree of competition. First, vertical integration by the major companies, which means that the firms which are dominant in one level of the industry are dominant in each of the other levels of activity. Second, a worldwide network of joint ventures and other intercorporate ties which must greatly moderate the inclinations of any one company to act in an aggressively independent manner.

Let us look at vertical integration in the domestic market. Here the effective control of the majors is extended well beyond the large percentages of total output which they themselves produce. Since the majors hold most of the lease acreage, much of the independent production comes on major company farmouts. It is true that this arrangement permits the survival of many independent producers who lack the financial resources to acquire their own lease blocks. But, almost without exception, farmout agreements give the leaseholder a perpetual call on the oil produced — in other words, with a farmout a major gives up some of his rights to produce oil from his lease, but he does not give up any right to control the disposition of that oil.

Next there is the crude oil pipeline system, which moves oil at a small fraction of the cost of any alternative mode of transportation, except for tankers and barges where waterways are available. Customarily, crude oil is sold at the wellhead to a gathering system or pipeline company. Thus, the company which controls the pipeline into a field controls the crude oil from the field. Very little crude moves on a true common carrier basis.

McLean and Haugh, in *The Growth of Integrated Oil Companies,* quote a 1948 Atlantic Refining Co. memorandum to the effect that owning the pipeline into a producing area is the next best thing to actually owning the reserves. John D. Rockefeller, Sr., knew this and acted on it much earlier.

Well, who owns the pipelines? From the pipeline companies which reported to the Interstate Commerce Commission in 1973, we find that 64 percent of the crude shipments in ICC lines originated in lines owned by the eight largest oil companies; more than 92 percent originated in lines controlled individually or jointly by the sixteen majors.

Next, there is the system of crude oil and product exchanges. While products are normally exchanged on a straight barrel-for-barrel basis, crude exchanges involve simultaneous purchases and sales — I suspect this was, until last year, in order to protect percentage depletion allowances, but it also helps to stabilize posted prices. Within the environment of vertically-integrated companies — and I would like to emphasize this condition — crude oil exchanges make economic sense. If I have a refinery at Point "A" and crude production at Point "B", while you have production at "A" and a refinery at "B", we can each save a lot of grief and cost by agreeing that I will buy your crude for my refinery and you will take my crude for your refinery. But what this does is to replace a competitive crude oil market with a system of bilateral and multilateral barter arrangements — a system from which non-integrated independent refiners may be easily excluded.

Perhaps I can make this clear by abstracting from a letter written by the president of Kerr-McGee to the Secretary of Interior in March 1973 — the letter is to protest an order to deliver government royalty oil from Kerr-McGee OCS leases to a small business refiner:

> *"Kerr-McGee exchanges or trades all of the oil produced by it in the Gulf of Mexico for oil to supply its refineries or for products to meet marketing requirements . . . When there still re-*

*mained surplus production in the United States . . . there was
no obligation to return crude to the various major companies
from whom we purchased. Beginning in 1972, . . . we faced increasing demands to sell these companies offsetting barrels of
Kerr-McGee's Louisiana production. This situation has continued to deteriorate until today we are literally unable to buy a
free barrel of oil from these large producers."*

In other words, there is not a workable crude oil market in the
United States today, nor has there been one for a long while. Money
won't buy domestic crude — unless you have your own oil to sell
in exchange.

Perhaps the principal goal of S. 2387 is to create just such a
crude market. Separated from their refining and marketing operations,
the major crude producers would have no choice but to offer their
oil on the market to any buyers, their own former affiliates or
independents. They would be faced in the market with aggressive
buyers, large and small, each of whom would be trying to get the
type and quantity of crude he needs at the best possible price. The
market would be competitive.

The second institutional factor I referred to earlier is the network
of joint ventures linking every one of the major companies. Within
the United States we find these principally in OCS production and
pipelines. Overseas, we find them at every level of the industry, from
production in Saudi Arabia to the Irish Refining Co. The industry
explains that joint ventures perform a risk-sharing function, which
may be very important to the companies involved. From the social
standpoint, however, we must recognize that each of these joint
ventures creates an interface among companies within which coopera-
tive behavior may provide a higher payoff than competitive behavior.

Let me give a dramatic example of the importance of joint ventures to particular companies. Many of the foreign assets of Texaco
and Standard of California are jointly held in the Cal-Tex Group of
companies. In 1975, according to *Petroleum Intelligence Weekly:*
"Caltex's profits accounted for 59.5% of Texaco's total profits, up
sharply from 37.3% and 26.5% in the two previous years. For
Socal, the reliance on Caltex also increased, but less sharply, rising
to 63.9%, from 59.3% and 40.5% of Socal's total profits in the
two previous years." The PIW figures do not include their dividends
from Aramco, which they share with Exxon and Mobil, or any of

their joint ventures with other companies. Now any disinterested observer would be justified in asking the question, "Are Texaco and Socal competitors or are they partners?" More generally, does the degree to which major oil companies profit from joint ventures with one another in some markets temper the aggressive spirit with which they would otherwise meet in those markets in which they are competitors?

Note that S. 2387 does not prohibit joint ventures. Producers would still be able to share the risks of exploration and development in frontier areas — although I would hope that Interior's present restrictions on OCS joint bidding would continue and the antitrust agencies would monitor joint ventures carefully to ensure that risk-sharing is not used as a vehicle for horizontal cartelization of markets. On the other hand, the separation of production from refining and marketing should do a great deal in itself to prevent such cartelization. At the same time, the required spinoff of pipeline operations to create a true common carrier system would prevent joint control of transportation from serving as a means of market control.

So far I have avoided the $64 question in any discussion of divestiture: What about OPEC? Isn't it true, as the oil companies themselves, and Gerald Parsky, William Johnson, and others have told us, that the integrated majors represent the best hope we have of negotiating with OPEC and holding down the world price of crude? My own feeling is that if this is our only hope, perhaps we'd better check the lifeboats.

Recall if you will the spring of 1973, when many of the OPEC governments secured 25 percent participation and announced that they would develop their own marketing capabilities as rapidly as possible, while President Nixon suspended quantitative controls on imports. Remember the large number of glorious plans for new independent refineries which were announced — and try to remember what happened to all but one or two of these plans.

What happened as the months passed, of course, is that after the first flush of enthusiasm over participation, the host governments came to the conclusion that the major integrated companies offered the best opportunity for marketing their oil in a manner which would avoid a break in OPEC prices. The door to independents appeared to close in Saudi Arabia, for example, last September when the

Saudis announced they would not enter any new contracts with private refiners. This appears to have been firmed up in the recent Florida meeting between Sheik Yamani and the Aramco owners. If the press reports are accurate, it was agreed that after nationalization of production the state company, Petromin, would retain only 5 percent of the nation's crude output to maintain the contracts it entered into in 1973 and 1974; 95 percent of the output would continue to be marketed by Exxon, Texaco, Socal, and Mobil.

This makes sense from the standpoint of OPEC. As M. A. Adelman told the Church Subcommittee last January: "The cartel governments use the multinational companies to maintain prices, limit production, and divide markets. This connection, I submit, is the most strategic element in the world oil market. The governments act in concert, the companies do not need to collude. The governments transfer oil to the companies at identical publicly announced prices. . . . The companies produce only what they can sell. So long as the governments are content to accept the market shares that result from the companies' sales efforts, the cartel holds."

I will go farther than Prof. Adelman and suggest that the OPEC nations are confident that the international majors will balance their liftings among the countries in such a way as to maintain satisfactory market shares for everyone. I see a parallel between OPEC's present situation and the debate in our own country in the early 1930's between the proponents of state prorationing and the proponents of federal prorationing. State prerationing won out, but it survived so long only because the large buyers were willing to preserve the posted price system by balancing their liftings among the prorationing states.

From the standpoint of the companies, this relationship with OPEC also makes sense. It's difficult to perceive any titanic struggle between the international majors and OPEC over the absolute level of OPEC prices through the past five years. Rather the negotiations have been over the marketing margins received by the companies, to their own profit, and retention of their ability to control the disposition of OPEC oil. They can pass price increases in OPEC oil on through their own refining and marketing outlets so long as they are assured that none of their independent competitors are getting that oil any cheaper than they are. If they are the sellers, they're safe in this respect.

How could divestiture weaken our posture as a consuming nation with respect to OPEC? The Aramco owners would still be producing

Saudi oil. The difference would be that they could no longer pass cost increases through their own integrated channels automatically.

Instead they would be faced by a group of the largest refining companies in the world, each of whose survival depends upon securing crude oil at the lowest possible price. These refiners would be negotiating worldwide, playing off one country against another. There is some price-cheating going on today among OPEC countries with respect to the small volume of their crude oil which moves outside the integrated company channels. With divestiture this area of competition would be vastly enlarged.

Now I am not talking about breaking OPEC or promising that crude prices will drop $5 a barrel the week after divestiture is ordered. What I am saying is that divestiture will help to impose a crude oil market discipline on OPEC oil prices, and that prices will be lower in 1980, or in 1985, than they would be without divestiture.

THE CASE AGAINST DIVESTITURE

Annon M. Card
Senior Vice President
Texaco Inc.

Two and a half years ago, Americans for the first time came to realize what it was like for their lives and our country to be dependent upon other countries for a necessary resource and to be subject to the political motives of those countries. Now, almost three years later, instead of making a concerted effort to change that situation, we are still talking about issues like divestiture and allowing ourselves to continue down the path of greater and greater dependency upon foreign oil and less and less ability to determine our own destiny.

Today, this nation is dependent on foreign sources for more than 40 percent of its petroleum, and, by 1980, this dependency will increase to 50 percent and might reach as high as 70 percent by 1990.

America must develop a realistic national energy policy that will assure our children and grandchildren adequate availability of energy. Divestiture does not offer the solution — rather, it intensifies the problem.

Let us examine the charges and arguments for divestiture.

Most advocates of divestiture claim that the petroleum industry is oligopolistic. However, no substantive evidence has been offered.

The Fact of the Matter Is . . .

That no one company or small group of companies in the oil industry has anything close to exclusive, and thereby obligopoly control. In no one sector of the industry does the largest company account for more than 10 percent. The four largest refiners, for example, control but 32 percent of the industry, as compared to the average of all other manufacturing industries wherein the top four companies control 40 percent.

The oil industry has traditionally been, and still is, one of the most competitive industries in this country. Today, there are more than 10,000 companies exploring for and producing crude oil and natural gas, 131 companies operating refineries, and approximately 300,000 gasoline retailers, of which about 95 percent are independent businessmen fully controlling their own businesses, such as establishing their own prices.

There is complete freedom of entry for new firms into every segment of the petroleum industry. For example, in the refining sector, twenty-one so-called "independent" companies have each constructed at least 50,000 BPD of new refining capacity since 1950. This includes Amerada-Hess which was not in the refining business in 1950 and now is one of the refining companies subject to divestiture. The only "grass roots" refinery currently under construction in the United States is being developed in Louisiana by a new entrant to the refining business.

Strong competition exists in gasoline marketing, and entry is easy. The non-major brand service stations have dramatically increased their share of the market in recent years.

In the producing area, statistics show that the number of companies both bidding for and acquiring offshore acreage has been rising sharply. Of equal significance, in underscoring the competitive nature of this business, is the fact that companies other than the eight largest have been increasing their proportion of offshore acreage in lease sales.

The Proponents Further Imply . . .

That there is collusion among the industry's major companies and, therefore, they should be broken up. Of course, if in fact there was collusion, it would be illegal under existing laws. Most commonly cited are the joint venture activities in bidding for oil exploration leases and the operation of pipelines. Let me state most emphatically that there is absolutely no foundation, nor has any evidence been offered, to support these allegations.

On the Contrary . . .

The facts show that the current bidding processes are intensely competitive. Joint bidding insures competition and results in the Government and thus the people benefitting through substantial bonus payments. The prime reason for exploratory joint ventures is to share the high risk and tremendous costs involved. If joint ventures were not allowed, small companies would probably be unable to participate. In addition, financing would be much more difficult if diversification of risk were not permitted.

Pipeline systems are basically common carriers and, as such, have for years fallen under the regulations of the Interstate Commerce Commission. As such, all shippers, large and small, are entitled to move their materials via these systems and pay tariffs which have been approved by the ICC. Pipeline financing is very dependent on the throughput guarantees which, except in very special circumstances, are provided by the owner companies who bear the risks. Joint ventures exist for the purpose of sharing tremendous investment costs; promoting efficiency; and obtaining the benefits of the "economies of scale."

Additionally, the Proponents Claim . . .

That the oil companies, being non-competitive, have been raising prices and thereby deriving huge profits.

The Fact of the Matter Is . . .

That the major oil companies make a total of about 2¢ per gallon on all products sold. Against that statistic, it is significant to note that federal, state, and municipal excise taxes on gasoline are approximately 12¢ per gallon.

For the past 10 years, through 1975, the high-risk petroleum industry has averaged profits of only 13.3 percent of net worth which compares to the average of all U.S. manufacturing industries of 12.8 percent. Unfortunately, this is not and will not be adequate to generate the huge amounts of capital required for reinvestment in order to supply ever-increasing demands for energy.

Concerning prices, during the 25 years prior to the embargo of 1973, the cost of gasoline increased by only 38 percent while, during this same period, the Consumer Price Index increased by 85 percent.

The most significant increases in petroleum prices have taken place since the embargo and the imposition of extensive U.S. federal controls and regulations on the entire petroleum industry. Increased costs caused by the actions of the various governments are the basis for the recent price escalations. It is clear that the oil companies are not responsible for these increases.

Those Who Would Dismember the Petroleum Industry Argue . . .

That breaking up the major oil companies would weaken the OPEC cartel.

One Should Understand That . . .

OPEC is an organization of sovereign producing nations which have banded together to further their own national interests. The major oil companies are not partners of OPEC. Rather, we are customers or service concessionaires of the separate nations that comprise OPEC, and, as such, the major U.S. oil companies have been able over the years to negotiate contracts and prices that have resulted in relatively low-cost energy for the American people.

Under proposed divestiture, the nations that make up OPEC would deal with the large foreign companies and small, weakened and disorganized United States companies, rather than with our country's present large, strong companies. This situation would favor the OPEC nations and foreign companies, and decrease our facility to effectively deal with OPEC.

So much for the claims of those wishing to destroy our oil industry; let us now consider the probable effects of the proposed legislation:

Less Energy Would Be Developed

It would take 10 to 20 years to complete divestiture. Extensive and protracted litigation would be inevitable. During this period, petroleum industry managements would be required to devote virtually full-time efforts to developing divestiture plans and implementing the divestiture program. The resulting uncertainty and chaos would make it virtually impossible for large sectors of industry to raise capital. The reduction in capital and incentives would discourage or delay new ventures with an inevitable loss of domestic production.

Greater Dependence On Imports

A reduction in domestic energy production would require offsetting imports from foreign sources. Such increased dependence on imports in turn would increase U.S. vulnerability to another embargo, which would endanger U.S. national security; weaken America's prestige; and impair our nation's freedom of action in world affairs.

Weaken The U.S. Economy

The reduction in U.S. petroleum activity would hurt every American. Clearly, reduced earning power by American petroleum workers would reduce purchases of goods and services throughout the economy. The offsetting petroleum imports would have an adverse effect on U.S. balance of payments. Foreign manufacturers would benefit rather than domestic industries.

Increased Costs To The American Consumer

Elimination of the efficiencies of integration would increase costs to the domestic industry. This, when combined with the weakening of the U.S. negotiating position relative to OPEC, makes it inevitable that the cost of energy will increase more than necessary to both individual and industrial consumers.

In summary, divestiture would disrupt energy supply; impair national security; stifle economic growth; and increase prices to both individual and industrial consumers.

What is needed is a realistic national energy policy implemented immediately; and a spirit of cooperation, not confrontation, between Congress, the Administration, industry, and the American people. If this can be accomplished, then the industry can do its traditional job for America by providing needed supplies of energy at reasonable prices.

The time for confrontation and debate has passed. We now must utilize, to the fullest extent, one of America's great industrial resources — our existing oil industry with its proven record of being able to get the job done.

By realistically facing our energy problems and cooperating with one another, we can lessen our dependency on foreign petroleum and strengthen our nation.

DISCUSSION

Robert Shepherd, Director
Office of Energy Policy & Programs, Department of Commerce

I think we would all agree that the large integrated and multinational oil companies have developed in response to changes in technology and marketing, as Professor Chandler has indicated. As a result of this they have achieved economies of scale, and by the very nature of that process they have built and extended considerable control over the products the market can handle.

Benefits and costs of divestiture

I think most people would also agree that divestiture would be a long, costly and torturous process. I cannot accept offhand the exact figures quoted by Mr. Card. I would want more detailed information, but I agree considerable litigation, some uncertainties, and even additional costs might result from divestiture.

I must ask, then, what are we going to gain by divestiture to make us go through this herculean effort? To my mind, none of the proponents, including Mr. Measday today, has made a convincing case that divestiture would produce products at a significantly lower price, or that divestiture would put the United States in an advantageous position vis-a-vis OPEC.

The proponents of divestiture have argued in large part that the very size of the industry — because petroleum affects all aspects of national life — must be reduced by legislation. But I do not think this case has been made.

One final point: it seems to me that the general trend in the industry is likely to be toward more concentration than less, and that this trend has been reinforced by an overlapping network of federal regulations. There may be in divestiture an objective to insure that smaller firms can operate, regardless of what happens to price. Perhaps this is an object worth pursuing, but that has not been stated in the legislation or by the proponents.

So I would toss to Mr. Measday my proposition that the proponents of divestiture have not made a case that the benefits would be worth the costs involved.

Walter Measday

I would like to briefly cite Maury Adelman's remark in the Washington Post on May 8, that the arguments for efficiencies in vertical integration are unfounded; and Ezra Solomon's statement, as Exxon's expert witness in the Wisconsin tax case, that the advantages of vertical integration are, in effect, trivial compared to other things, because you don't have vertical integration between production and refining in the usual sense.

The type of competition we envision is a little bit different, perhaps, from what free enterprise means to a businessman. We've had to have these regulations to keep the independent sector alive in the past few years.

Regulation a necessity

Now, I'm not making a plea for the independent sector here. I'm saying that if we have a crude oil market available to the independent refiners and to the majors alike then we let the battle be fought on the basis of efficiency. I don't want to save any independent refiner just because he's an independent. But I think if he has demonstrated his efficiency over the years, and if he has access to the crude oil market he should continue to survive on that basis. I think divestiture is a better way to go than greater federal regulation, which I see as the alternative. FEA is not going to disappear 39 months from now unless we do something like restructure the industry to preserve the competitive market.

Annon Card, Senior Vice President, Texaco Inc.

I would like to make three points. The first point involves the direct question which really wasn't answered. What are the benefits that have been anticipated in divesture? Dr. Measday's answer is a typical proponent response. Proponents cannot demonstrate that there will be lower prices and in fact they are 180 degrees off the mark. There will be higher prices under divestiture.

Second, they have not demonstrated that there will be increased energy supply. The best hope is to have the industry doing its job and over a longer period of time become less dependent upon imports.

The third point involves Mr. Measday's contention that some companies have already arranged for divestiture in their organizations. This is totally untrue. There has been no divestiture or any arrangements for it under the operations and reorganizations that are normal occurrences and which take place from year-to-year under normal activities.

William Smith, Business Correspondent, The New York Times

Theoretically at least, a reporter's job is not to make statements but to hopefully ask penetrating questions. No penetrating questions are promised. I have generally been puzzled by the ascendence of the divestiture issue. Because of the potential consequences of this question, it is important to gain as full an understanding of its genesis as possible.

Contradictions in arguments

In keeping with the theoretic role of a reporter, I have some questions I would like to ask — maybe not on major issues but on things that tickle me, because there is a little bit of the non-sequitur that happens when people speak.

First, I would like to ask Mr. Measday if Maury Adelman, who has been quoted here, is a proponent of divestiture?

Walter Measday

No, Maury Adelman is against divestiture.

William Smith

I just think that's interesting. (laughter)

Walter Measday

He'd much prefer his own coupon auction plan for purchasing imports, which I think would be incredibly more complicated than divestiture.

William Smith

Now Mr. Card it strikes me how the industry, in its opposition to divestiture, keeps saying how inefficient it's going to be, that it will force duplication of jobs and efforts. Then in your presentation you said there would be a loss of a million jobs. How do you explain this apparent contradiction?

Employment effects

Annon Card

Well, first of all, there is a direct correlation between available energy, the Gross National Product, and unemployment. Historically, prior to 1975, it's taken about 1.2% increase in energy to support about a 1% increase in GNP. Even without divestiture, there's going to be increased dependence on imports, and with divestiture this increase would certainly accelerate. Who is to say that those imports would be available? If they are not available, jobs might be lost and an immediate, chaotic situation would develop in the petroleum industry. I already pointed out it would take management 10 to 20 years to try to cope with any kind of planning, and put the industry back together.

This is a tremendous and complex business. My own company supplies some 34,000 retail outlets, hundreds of thousands of customers in 50 states from over 200 salary-operated terminal points and about that many wholesale operations. To get the product there when you need it is a tremendously large and complex operation. In our case it took some 74 years to put this together as an efficient way of doing business. I pointed out that during the 25 years prior to the embargo, gasoline prices increased 38%. At the same time, the consumer price index increased 85%. We claim this is pretty good efficiency. Now they're proposing to rip this apart. Given problems of supply from chaos in the industry, there will be many people in the country — many hundreds of thousands, even millions — who would not be employed if the energy were not there to support the industry.

William Smith

I want to just get back quickly to Mr. Measday. Now if we

Divestiture of foreign companies

break up American oil companies, what are you going to do about someone like Royal Dutch Shell and British Petroleum? And how is this going to affect our posture overseas?

Walter Measday

We would certainly have some diplomatic problems there. I would suspect that it might be easy enough for Shell to spin-off Shell U.S. I think this is what they would do. BP would be faced with a more difficult problem since they are losing so many of their overseas reserves now. They'd have to decide whether to divest themselves worldwide or to give up the North Slope. I don't know what they've got up there, but BP probably has about 10 billion barrels of crude reserves in the North Slope. They'd be faced with a tough problem. If you don't like the laws, you do business someplace else. I doubt, for example, if Aramco is hiring very many Jewish employees to work in Saudi Arabia. Companies run into this all the time. If you want to do business in a country, you obey the laws of the country in which you're doing business.

Lee Richardson, President, Consumer Federation of America

Mr. Card, you've been using the terms efficiency and inefficiency as they relate to the advantages of scale economies. Isn't

Efficiency of retail system

it true though that the marketing operations of the major integrated companies are generally considered very inefficient and, in effect, are subsidized? It's been said that we've got to reduce retail outlets by a thousand gasoline stations. Isn't the retail system of the majors a case of inefficiency?

Annon Card

Well, the service station program that has been developed over the past several decades has provided the consumer with the high quality products at reasonable prices that they need, wherever they need it, and when they need it. This performance to me is the definiiton of efficiency. It has been true in marketing throughout the petroleum industry, and I think this has been true regardless of the companies involved.

We've talked about the distribution of crude oil and products in the 50 states. To get product there requires efficiency

Industry efficiency

and planning. You have to have the deliveries there on time whether through pipelines, trucks, or tankers. You have to adjust refinery runs to be sure that you have the right products in the season that they're demanded. As you probably know, refinery runs change in the spring for the gasoline needed during the peak driving season. In the winter they change over to the manufacture of heating oils and other seasonal products. All this is efficiency.

We believe that all the components have been managed in a highly efficient fashion to bring to the American people their petroleum needs at the reasonable prices.

Lee Richardson

Is the increased market share of the so-called independents at retail level evidence of this same efficiency?

Annon Card

The increased market share of the so-called non-branded retailers is actually an indication of the competitive nature of this business. The proponents of divestiture claim that it is not competitive, that by breaking up the oil companies, it will be competitive. The oil industry is as competitive today, as any other industry, as the evidence on the ease of entry, and the independents' share of the market, indicates. All these factors indicate intense competition.

Lee Richardson

I'd like to see more evidence from either speaker on just what the economics of scale are. If we had the economies of scale as we hear from AT&T, ultimately we should have seen this industry evolve into perhaps one large, giant, "efficient" integrated company. If such efficiencies did exist in all areas, AT&T would have continued to grow. These market share figures wouldn't appear as competitive as they do today. Yet we have not seen this kind of growth. Just where are the economies of scale?

Economies of scale

Annon Card

Let me give you an example of this. You cannot in most circumstances have a 50,000 barrel a day refinery today or even a 100,000 barrel a day refinery — a sophisticated refinery — and have the economies of scale that are necessary. A 200,000 barrel a day sophisticated refinery can cost as much as a billion dollars, depending on the kind of facilities that you have and the uses you expect to put the facilities to. The cost of a 100,000 barrel a day refinery is not that much less, but with the smaller refinery you'll have no way you can recover the investment over a reasonable period of time.

Pipelines provide another example. The economy of scale between a 12-inch and a 36-inch pipeline is tremendous but the cost differential isn't that great — thus the economy of scale. Comparison of the cost and capacity of a super tanker to a T-2 tanker also demonstrates economies of scale. These economies are one of the principal reasons why the cost of energy in this country has been maintained at what we think are reasonable prices.

Walter Measday

Let me just say on economies of scale — certainly in refining — your long-run average cost curve seems to go down to about 200,000 barrels a day, but its costs are not much greater at 100,000 or 150,000 barrels a day. With geographically dispersed refineries much smaller capacities might be suitable. Texaco, for example, has about a dozen refineries in the United States. One of them is over 400,000 barrels a day, one is 140,000 barrels a day and 10 of them go from about 84,000 down to 17,000 barrels. I assume these are still operating efficiently on a locational basis. When you consider the costs of moving products from a huge refinery to a distant marketing area, perhaps a smaller refinery in the marketing area can operate successfully. I'm not sure there are any multiplant economies of scale in operating a large number of refineries. I can't see, for example, where Exxon get economies above those of a single refinery from operating refineries in Linden, New Jersey; Baytown, Texas; and Benecia, California, production and refining, simply because it's not a continuous process. You don't move your own crude to your own refineries necessarily. We asked Exxon this, and Exxon said they cannot tell us how much of their own crude they run in their refineries. They buy and they sell and they trade; they get a pool of crude and it goes to all refineries. But how much of it is theirs they don't know. Furthermore, I doubt if Mr. Card could tell us how much of Texaco's own crude is run in Texaco's own refineries. I don't see vertical efficiencies there between crude and refining. And I think it is easy to exaggerate the economies of scale at the refining level.

Albert J. Anton, Jr., Partner, Carl H. Pforzheimer & Co.

During the last several years, the Federal Trade Commission has been pressing an antitrust case known as the "Exxon Case" against the eight largest oil companies. Thus far they seem unable to develop a case. Perhaps the antitrust laws as they're written don't do what the FTC is trying to do. If the companies are doing something antisocial, perhaps we need a new review of the antitrust laws. My question, to perhaps both of our speakers, is if monopolistic elements do exist in our economy, shouldn't they be addressed through existing or revised antitrust laws? And furthermore, shouldn't they be addressed fairly and equitably to all sectors of the economy — to business, finance, labor (which is basically exempt from antitrust), and to government policy itself — not just the oil industry.

Improved antitrust laws

Annon Card

Well, I've said in my prepared talk that such charges of monopolistic practices, etc, are totally without substance. No cases to date indicate that such charges are valid. So I say that no such practices are in effect.

Walter Measday

Even though S. 1284 (The Antitrust Improvements Act of 1976) would improve antitrust, still the problem remains, that major antitrust cases take too darn long. For example, the IBM case was filed in January, 1969. It finally went to trial just last summer. The trial will go on for another year, and with appeals a final decision may come sometime in the early 1980's. The Exxon case was filed in July, 1973, and the FTC is still in the discovery process. In cases which have been handled — and it may be a failure of antitrust law or antitrust resources — the antitrust agencies have acted upon minor conduct issues without making any changes.

Deficiencies of antitrust

I'll give you an example: the case of a supplier requiring his retailers to carry a certain line of tires, batteries, and accessories. The first case that I know of was about 1940. There have probably been 40 of these TBA (tires, batteries and accessories) cases since then, and in everyone of them the suppliers are told that they can't tie the station lease to TBA's. But the companies go right on doing it. The cases keep coming up.

In another case, the Socal case of 1950-51, the courts ruled that you cannot require your dealers to carry only your brand of gasoline. Yet at hearings on gasoline marketing we asked a number of majors, Mr. Staub of Shell, for example, if Shell required its dealers to carry only Shell gasoline? And he said absolutely not, as long as they don't sell it under a Shell label, they can carry somebody else's gas. Asked how many of your dealers do it, he said none of them. He said they're all convinced of the superiority of the Shell product. I submit that at least the Shell dealer I know might be willing to sell other brands of gasoline, although I don't know how long he would be in business if he added a Texaco pump in a Shell station.

The antitrust people have only gone after these fringe aspects. The last real structural case we had was the Standard Oil case in 1911. The Committee thinks that we've got to deal with industry structure. And the only way we can correct that structure within a reasonable period of time is actual congressional legislation.

Albert J. Anton, Jr.

I'm not denying that perhaps legislation is required. I just wonder if in the structure of labor unions and in industries other than oil there aren't inflationary elements which a review of the antitrust laws could address. Such a step would eliminate the need for this bill and put the whole economy on a fair and even basis.

Broad review of antitrust laws necessary

Walter Measday

We've given these questions a lot of thought and come to the conclusion that divestiture is the way to go for the petroleum industry. I will also point out that Senator Philip Hart has put in the Industrial Reorganization Bill which deals with other sectors of the economy.

Albert J. Anton, Jr.

Does it comment at all on labor?

Walter Measday

Yes, we've gotten a lot of flak on that. The Industrial Relations Commissioner is required to examine labor contracts and practices and would have to report to Congress any of those who appear to have anti-competitive effects for congressional action.

Annon Card

I'd like to comment on the Committee's consideration of alternatives and its decision that divestiture would be best for this country. In making this decision they are gambling recklessly with the welfare of this nation, its national security, the employment of the American people, the very basic resources of the petroleum industry and the welfare of all of us.

3 HORIZONTAL DIVESTITURE: OPPOSING VIEWS

THE IMPORTANCE OF COMPETITION AND HORIZONTAL DIVESTITURE

Floyd Haskell
Senator, State of Colorado

One of the major questions we are asking today is this: Is it advisable for the major energy sources of the nation to be under one roof? I think the real question probably is how do we get the biggest bang for the buck? That is a reasonably inelegant way of paraphrasing what the Supreme Court said in the Standard Oil Case of 1911.

It was, of course, the case that broke up the Standard Oil Trust, and the courts said "the unrestrained interaction of competitive forces will yield the best allocation of our economic resources, the lowest prices, the highest quality, and the greatest material product." I think this is our basic economic belief in this country — that competition does all these things. That basically is the American capitalistic system; there are other countries that don't believe that. In this country we do.

I think what we are looking at here is how do we maximize the production of all our energy sources? What institutional forms should control our energy resources? Should they all be produced and owned by the government? Should each company deal with only one resource? Or should the various types of energy resources be held by one company?

I opt for each company working in only one resource; I think competition is served better in that way. Let's look at a few figures which show what I consider an alarming acquisition by the major oil companies of the alternate energy sources, although I'm sure they consider it good. From a public policy viewpoint, I can't agree. According to Bureau of Mines figures, the oil companies own 57.3% of the total coal reserves under lease in the United States, including those on both private and public land. As far as uranium goes — I would consider uranium, coal, oil and gas the major energy sources — the oil companies own 65% of $8 reserves.

The oil companies own 69% of uranium milling capacity (which of course is extremely important — roughly equivalent to refining, although there are two steps in getting the end product). This is a trend which has been accelerating over the past decade. The chairman of the Tennessee Valley Authority, which is probably the major purchaser of coal in the United States, rather dramatically called attention to the changed situation in the ownership of energy sources. He says, "In 1960 only one of TVA's major suppliers was *not* an independent coal company; in recent years, of the ten major suppliers (that supply more than 70% of TVA's coal) only one was an independent. Seven of the current major suppliers are controlled either by large oil companies or conglomerates engaged in developing or marketing oil and gas."

So the change is dramatic, the trend continues, and I suppose what we have to ask ourselves as a nation is: Will that help production? Will it promote competition? Because it is our national belief that competition results in the lowest prices and the highest quality. Therefore we always have to ask ourselves if a given situation promotes competition?

I think in this equation it is important to see the interaction of fuel prices. Do prices tend to follow each other, or do fuel prices vary? As you would expect, fuel prices follow each other in extraordinarily close parallel. For example, from 1972 through 1975 the price of oil increased 136%, natural gas 134%, coal 145%.

Basically, of course, whether you have a competitive or non-competitive situation, you would expect fuel prices to follow each other. And if you have a shortage, you would expect them to follow each other up. So one of the questions, given the current state of demand for energy is whether it is ever possible to, in effect, have an oversupply and therefore competition on the downside? That is the question we ought to ask ourselves.

One figure, I think, is highly significant in answering that question. Using Bureau of Mines statistics, the extractable coal reserves — not the total coal reserves, but the recoverable reserves in this country — at the present consumption rate of 600 million tons a year, would last us 453 years. So we've got lots of coal. And, of course, it is entirely possible — and I hope probable — that we will increase our coal production so that it is truly competitive with oil and gas, so that there will be pressure on the downside, an oversupply.

If coal, oil and gas, and uranium are all in the same hands, is there going to be competition? Suppose I own all three of them. If I see that market interaction is going to mean that all three are going to go down if I lower the price of one, am I going to lower the price? Common sense would tell me no. I'd be a darn fool to do so. And I would suggest that, if all fuels are in the hands of the oil industry, the oil industry will, if faced with an oversupply, curtail production rather than cut price.

This, of course, has been the classic reaction of the industry. Again it is a self-protective measure; I would probably do the same if I could get away with it. It is self-protection and they have done it — witness the Texas Railroad Commission, which of course has been doing it for over a generation. Recently, it did away with prorationing because of the situation. However, prices have softened in natural gas, and the other day the *Oil Daily* said the Texas Railroad Commission would consider hearings on reinstituting prorationing.

I would think it would be in the best interests of our nation that these various energy sources be in the hands of true competitiors and under different corporate umbrellas. This is not an unusual request. We have legislated over the years that airlines can't hold aircraft companies, bus lines can't own railroads, and railroads can't own airlines. Where there is natural competition between different forces, we figure it best that there be a separation of ownership. The same thing is true, of course, with banks going back to the thirties: the bank has to decide whether it is in the investment business or the banking business. I think this is a classic response in this country — to have distribution rather than consolidation.

I'll quote — since I suppose everybody has to quote somebody — Thomas Jefferson. I think this quote is applicable not only to government but to economics. He said, "It is not by the consolidation or concentration of power but by its distribution that good government is effected." And ladies and gentlemen, I would submit to you that it is in the national interest that these energy sources be distributed.

THE DEMAND FOR
ENERGY AND
HORIZONTAL INTEGRATION

W. T. Slick, Jr.
Senior Vice President
Exxon Company, U.S.A.

Any meaningful discussion on divestiture must take place with the realization that the United States will, of necessity, continue to rely heavily on foreign oil to meet its energy needs for many years. Imports of roughly 50% of oil requirements will be needed by 1980. The question we should be paying closest attention to, therefore, is *how* can America accelerate development of *all* potential domestic energy sources and minimize this growing reliance upon uncertain foreign sources.

I don't pretend to have any easy or guaranteed answers. It's important, though, that we consider the consequences of wrong answers. To err and fall further short of domestic supply would compound current energy problems which have their roots in previous short-sighted legislative and regulatory policies.

There are few decisions that would be more devastating to the nation's future energy and economic growth than choosing to dismember America's most efficient and productive oil companies or to deny them the right to compete in other energy businesses.

Senator Haskell and others, however, seem to be concerned that large oil companies are attempting to monopolize all forms of energy and thus be in a position to regulate development and production, control prices and reap exorbitant profits. There is simply no basis for such concerns. No amount of conjecture can deny these three facts: first, companies involved in supplying each form of energy are highly competitive; second, entry of oil companies into other energy industries has increased competition, increased production and expanded research and development within those fields; third, legislation that would forbid such diversification would reduce competition.

I realize no one is surprised to hear such statements from a representative of a major oil company, but let me expand these points — as best I can in the few minutes available — to tell you why they can be said with such certainty.

I'll begin with the fact that the nation's energy industries are highly competitive.

This is clearly established by examining the concentration ratios for total energy and individually for oil, gas and coal. In each case, these ratios are low, both in terms of commonly accepted absolute standards, and relatively, as measured against the average for U.S. manufacturing. For example, according to the latest available data, the top four companies accounted for just 21% of total energy production, compared to an average of 39% for all manufacturing. The top eight energy firms had only 34% of production, versus an average 60% in manufacturing.

Equally important, the top firms in supplying one form of energy are not the leaders in supplying others. In fact, if you draw up lists of the top producers in oil and gas, in coal and in uranium, you have to list fourteen to find one company that makes all three lists. And that company is dead last on the oil and gas production list. Simply put, oil companies do not dominate any one fuel industry, nor all of them combined.

On my second point, the record shows that entry by oil companies into other energy industries has increased production and expanded research and development, thus increasing competition. This has been true whether an oil company began a new operation or acquired an existing firm.

In uranium, for example, Exxon started its exploration efforts in 1969. Since then we have made uranium reserve discoveries in Texas and Wyoming. These are new reserves that have been added to the nation's domestic energy supply — reserves which may not have been found without our introduction of new, added competition.

In the same way, the oil companies and their affiliates operating in coal — far from restricting production — have in the aggregate increased coal output faster than the industry average.

Such increased competition and production, however, does not inevitably lead to domination of an industry. Exxon, which began its coal operations from scratch about seven years ago, produced nearly three million tons last year and our development plans could result in production of 50 million tons by 1985. Even so, our potential output in 1985 would be only about 5% of projected total U.S. coal production in that year.

I can't predict who will account for the other 95 percent, but I do know the opportunity for others to compete in the coal business is wide open. Ample proof again can be found in the facts: over 45% of current recoverable U.S. coal reserves are unleased. About two-thirds of those coal reserves which are leased are owned by non-oil companies. In fact, oil companies now own only 19% of the nation's recoverable reserves. And the percentage of leased reserves held by oil firms has actually declined in the past year or so.

There will always be many potential non-oil entrants in other fuels. After all, energy supplies such as coal and uranium are purchased predominantly by large firms such as utilities or steel companies. These are sophisticated buyers who already are in, or are capable of entering, the coal or uranium production business themselves.

Moreover, when oil companies enter other energy industries such as coal, they compete against each other, as well as against non-oil firms. The result being that all new competitors — whether oil or non-oil companies — increase competition in individual fuel industries and enhance the opportunity for competition between fuels in the long run.

Entry by anyone into the coal, uranium, synthetics or other energy business is limited only by existing regulatory barriers and by the number of companies willing and able to assemble the necessary skills and technology and to assume the investment risk.

Many business similarities among the energy industries make oil company diversification into other fuel businesses a logical development. One important similarity is the fact that coal, shale oil, and tar sands are — like crude oils — forms of hydrocarbons. Exxon and other oil companies have spent more than 50 years studying and working with hydrocarbon technology. The need and incentive to expand petroleum technology to the related technologies of other hydrocarbons are obvious.

In the case of uranium, more than 90% of the nation's uranium reserves have been found in geological settings very similar to those of petroleum. In fact, two Exxon uranium reserve discoveries arose directly from our petroleum activity. Uranium exploration efforts are a logical extension of an oil company's expertise.

I could go on: other similarities among energy industries include high technology content, land and resource management skills and expertise in handling large-scale, capital intensive ventures requiring long lead times. The point is that product diversification resulting from the expanding application of basic capabilities is a sensible expression of the competitive process — one that is common in almost all U.S. industries. The result of such diversification can only be beneficial to the nation.

While oil companies introduce many forms of competition into other energy industries, it is the competition of ideas that is perhaps most important. The petroleum industry has an enviable record of technological advances and innovations. And it has brought that same desire and ability to innovate to other energy industries. Since 1964, 49 patents have been granted to oil companies in synthetic fuels technology alone.

If diversification had been forbidden 10 years ago, would these and other such breakthroughs have occurred? Perhaps, but it's difficult to imagine who would have been able and willing to undertake such research and development projects considering the costs, risks and uncertainties associated with them. Oil companies have done so because costly and risky R & D efforts have been a way of life in the industry since its inception, and because our continued business existence depends upon timely development of alternative energy sources to replace rapidly declining natural reserves.

There are many other parts of the energy spectrum in which oil companies — and other energy companies — are involved: solar, geothermal, nuclear fuel manufacturing and processing. In each case there is logic to support these activities and, therefore, logic against restrictive legislation.

Denying oil companies the right to enter other energy businesses would deny the nation the benefits of *our* ideas. Ideas will be generated by others. But they won't be the same ideas, and the potential contribution of ideas from any quarter — including the petroleum industry —is too valuable to lose.

The so-called horizontal divestiture bill is entitled the "Interfuel Competition Act of 1975." It would be more appropriately named the "Energy Competition *Reduction* Act," for that is what will result — less, not more, competition.

The United States already has effective anti-trust laws which provide sufficient safeguards against possible anti-competitive business practices. Adding legislation that sets arbitrary limits on who may compete in which energy industries would not protect or increase competition — just the opposite. Such legislation would actually protect firms in petroleum, coal, nuclear, and other energy industries from present and future competitors. It would also discourage existing coal or uranium companies from entering the oil and gas business and discourage non-energy companies from entering into any fuel sector.

The United States is fortunate in being endowed with a tremendous energy potential. It is fortunate to have a variety of competitors working within the world's most productive economic system.

But the United States can realize its energy potential only if artificial restraints on potential producers of all forms of energy, whoever they may be, are not enacted.

DISCUSSION

Albert J. Anton, Jr., Partner, Carl H. Pforzheimer & Co.

If the horizontal divestiture bill is passed restricting the entrance of oil companies into other energy sources, I wonder what these companies will do as their reserves of oil and gas run out. Oil and gas reserves-to-production ratios in the United States are much lower than any other energy source, such as coal or uranium. These companies obviously want to sustain themselves, and it seems to me that as their supplies disappear,

Future of oil companies in energy business

they will either liquidate themselves or will have to invest their cash flow in businesses outside the energy area. One would expect the oil companies to be logical channels of investment into new energy areas, and I would think, then, that horizontal divestiture might help defeat the whole objective of domestic energy self-sufficiency.

Senator Floyd Haskel (D-Colorado)

Importance of
interfuel
competition

Well, in the first place our proposal is not meant to benefit the oil companies; it is meant to be a benefit for the nation. It seems to me that intense competition is desirable. It also seems to me that the proposition I espouse will foster that competition. I cannot see a company allowing, for example, a coal subsidiary to compete with an oil subsidiary; it doesn't make sense. As a matter of fact, Mr. Hardesty of Continental Oil answered the question — "would you permit your coal subsidiary to undersell your oil industry?" with the answer: "No, sir under no circumstances. We're not going to play one source of energy against another." I wouldn't either. If we believe in competition, then the best way to foster competition is to be sure that a group of companies who have coal, a group who have uranium and another group of companies who have oil and gas are trying to underbid each other. Just because we might run out of oil and gas and perhaps at that time an oil company will have to go into some other line of business doesn't mean we should forget about competition. I don't know whether you can say with certainty that we are going to run out of oil and gas. Certainly the decline curve is there, but I don't know whether it is a definite proposition. In any event, I think that energy sources under different people promotes competition, and I think that is in the national interest.

Albert J. Anton, Jr.

Doesn't your view of Continental Oil not underselling itself avoid the fact that there are other energy companies competing with Continental Oil? Some of these companies, for example, produce coal in only one area of the country, and uranium in another. I can't see how they even compete with themselves; their vested interest is to have the uranium division compete with other producers of uranium, and to have their coal operations develop satisfactorily in the coal market. Kerr-McGee, which is one of the few companies in all these forms of energy, has been extending its financial resources to develop all these different forms of energy. The company is by no means a leader in two of these resources.

Senator Floyd Haskell

As I mentioned, historically fuel prices follow each other. If there should be an oversupply of any type of fuel — and I think it is entirely possible in this nation there will be — do you think that I, as an owner of all three fuels, would be about to reduce the price of one fuel when I know this is going to make other firms follow and make other fuel prices drop? Such a move would be pretty dumb, from my selfish viewpoint.

Energy price and production adjustments

It seems to me that I would go back to what is the oil industry's classic response to an oversupply, and that is to cut back production. I mean, that's the more logical response — none of us can sit here and say that is exactly what would happen, but to me it has more logic.

Albert J. Anton, Jr.

If I were Kerr-McGee, the largest uranium company, and perhaps the 30th largest oil company and the 20th largest coal company, I might indeed cut the price of uranium if it were in my interest to do so. In contrast, at Continental Oil the coal division has become such an important part of the company, that I think the economics of coal — not oil or uranium — would determine its policy. The interests of these companies are not parallel.

Senator Floyd Haskell

Maybe that is what makes horse races.

William T. Slick, Jr., Senior Vice President, Exxon Company, U.S.A.

The subject of substitutability and cross-elasticity is pretty complicated and one I don't intend to get into. Senator Haskell's concern seems to center on the notion that each one of these fuels would be controlled by a small handful of the same companies. I obviously don't agree with that. He has made the point that it would be in the interest of a company to cut production in one fuel to increase the price of it. The one comment

Impact of reducing production to raise prices

I would make is that if you have 5% of the coal business and 5% of the oil business and if you cut the production of coal so it increases its price, you therefore will sell more of your oil. But, first you lose 100% of what you cut back in the way of production. Second, you are wildly assuming that everyone in that industry is going to go along. If they do that, it's against the law. And third, for every one ton of coal you give up you get no revenue; and you only have your 5% of the oil industry to increase production and offset that loss. So it doesn't make sense for somebody in one fuel industry, under any circumstances that I could envision, to cut production in that industry in an effort to drive up prices in order to sell more in the other industry.

Senator Floyd Haskell

Let me, Mr. Slick, point out the history of the Texas Railroad Commission that I am sure you are very familiar with. Just recently, as I mentioned, apparently there was an oversupply of gas, and a softening of gas prices in Texas. *Oil Daily* reports that the Texas Railroad Commission is going to have a hearing as to whether or not to reinstitute proration. All proration does is cut production.

Prorationing and energy prices

William T. Slick, Jr.

Senator, two points: I'm familiar with the history of the Texas Railroad Commission, as I am sure you are. Texas Railroad Commission has been holding hearings on proration every month since I got in the business. They held them last month, the month before, and they'll have them the next month, and conjecture on the part of the *Oil Daily* on why they are holding hearings ignores the fact that they hold them every month — it's their job to hold hearings. Secondly, the Texas Railroad Commission's regulatory practices in the oil and gas conservation area have been tested in the Supreme Court of the United States and the Supreme Court says they are consistent with the laws of this land.

Senator Floyd Haskell

Yes, I would say though, Mr. Slick, that *Oil Daily,* which is your industry's publication, seems to think it was newsworthy.

William T. Slick, Jr.

It's Mr. Franshier's publication; not my industry publication.

Senator Floyd Haskell

And second as far as being the law of the land, we're discussing what the law of the land ought to be. Not what it is.

William T. Slick, Jr.

Touché.

Lee Richardson, President, Consumer Federation of America

I've heard over the last couple of years how we need to deregulate the price of oil and gas so there would be an adequate incentive to allow us to find all the domestic oil and gas that we have in the United States. Now I hear about the tremendous opportunities Exxon is finding in increasing coal from 3 million to 85 million tons of production. I just wonder if really you have a motive here of just finding places to put money that you made so much of (that, of course, doesn't show up in your return on equity figures as reported but nonetheless there seems to be quite a cash flow here). It is not going into oil and gas; you are now expanding into other industries. So, Mr. Slick, in spite of what appears to be, to a non-technical person, a certain shared technology in both exploration and chemistry of energy I wonder if you're really just trying to place extra dollars, as in Montgomery Ward.

Petroleum industry investments outside the energy industry

William T. Slick, Jr.

Your comment says that the profits don't show up in the balance sheet. You know, they are not accumulating in a sack someplace, and financial reporting by publicly held companies is regulated by the Securities and Exchange Commission. So I don't know where you think these profits are accumulating or how they can go unreported. Maybe I'm overly defensive when I react to your comment about trying to find a place to put your money. Well, that's what businessmen spend all their lives doing, trying to find good places to invest their money. We're in the energy business. We're trying to find opportunities to invest money in the energy business.

One of the factors retarding investment in new domestic oil and gas is the rate at which the federal government decides to have lease sales. Even then you don't know whether you will be successful in them. We think it is in our interest and it is in the interest of the country to diversify into other energy fields. And yes, we're investing money in the coal business; we reported it in our annual reports and we're quite proud of it.

Lee Richardson

Mr. Slick, you mentioned this 14th ranked company that was the highest ranked oil company in the three energy areas. What does the picture look like by 1985? I would think quite different in view of the statistics the Senator quoted about the industry, control of reserves, and milling capacity.

Future industry concentration

William T. Slick, Jr.

Well, you're asking me to speculate on what other people are going to do; I haven't the foggiest idea and I said so in my prepared statement. Insofar as the milling capacity numbers are concerned, nuclear power today represents less than two percent of the energy consumed in this country; by anybody's definition it is an infant industry. I submit to you that concentration ratios in the uranium industry are going to change markedly. I can see no way the leading company in the industry can maintain that position, and I think you are going to see a lessening of concentration in uranium mining, milling and exploration rather than an increase. There are some eighty companies in this country exploring for uranium. They may not all be successful, but some of them will be.

William Smith, Financial and Business Correspondent, The New York Times

Mr. Slick, suppose for the sake of this question we grant all your contentions: can you in the future envision the possibility of control over price? It might not be done deliberately, but would occur as the industry grows, as the strength and technological skills of the oil industry develop into a situation where you would not have inter-fuel price competition.

The possibility of noncompetitive prices due to concentration

I should hark back to Professor Chandler's comment, which I think was the most essential thus far today, that there should be new ways to control and improve the situation without going back to the answers of our fathers and grandfathers. Would it not be better to admit control of price could happen, then stop further investment, and provide a stable environment for other corporations to go into these areas? The oil industry often says that a stable investment situation is the key to getting new capital. Wouldn't it be better to stop now, get Congress off your back in this area, and let everyone proceed with developing the alternate energy sources the country needs?

William T. Slick, Jr.

You have asked two questions. Can I, under any circumstances by the end of the century visualize the energy industry getting concentrated to the point that somebody can control prices? No. Would it be better to forestall that possibility, which I can't visualize, and pass this law today? Again, no.

One of the things we're going to have to decide in this country is whether we want to develop energy in the fastest way possible, to meet the needs of this country in a manner consistent with the framework we have. I was struck by the same comment that Dr. Chandler made this morning, but another occurred to me about "new ways." Have you read the FEA regulations lately? They have found more new ways to regulate oil industry in this country than you can imagine.

William Smith

Again, I can still envision the possibility of a non-competitive price structure. And I, again following Professor Chandler's comment, just wonder whether the industry should look at controlling itself.

William T. Slick, Jr.

There has been a great preoccupation with 1911 today. I can only say to you that prior to 1911 there was a virtual monopoly, and I think that Dr. Chandler would probably come close to agreeing that one company controlled eighty-odd percent of the refining and marketing, and almost one hundred percent of the transportation. What is the relevance of that today? Since 1911 — sixty-five years ago — the piece of the company that I am with has been running hard; but today we've got eleven percent of the oil business and eight percent in another. If I could project that another sixty-five years down the road, I couldn't really see anything changing.

The preoccupation with 1911

4 THE UNITED STATES TREASURY ANALYSIS: THE EFFECTS OF DIVESTITURE

Gerald L. Parsky
Assistant Secretary
Of The Treasury

It is indeed a pleasure to have the opportunity to share with this gathering of experts my thoughts on the subject of energy industry divestiture. The timing of your conference is particularly appropriate because of the current public and Congressional debate on the issue.

Several weeks ago we convened a special task force within the Treasury to take an objective look at some of the potential consequences of various divestiture proposals. Our analysis is almost completed, and I would like to share with you today some of the preliminary conclusions. We have purposely concentrated on the financial and economic effects of the divestiture issue, particularly the likely effects on capital formation in the energy industry. We have also examined the effects on the structure and operations of the domestic and international energy industry as well as considered some of the legal aspects of divestiture.

I would like to offer you our views on each of these subjects; but in the end, I believe the most important consideration should be the effect of divestiture on the U.S. energy objectives. As I will discuss later in some detail, we have concluded that divestiture would seriously hinder the achievement of our national energy goals, while not fulfilling the objectives claimed by proponents of divestiture.

Divestiture and Our Free Enterprise System

In order to appreciate fully the consequences of divestiture, it is important to view the proposed legislative actions as part of a general policy choice that faces all of us today. Although legislated divestiture of the oil industry is not a new idea, I believe the present level of support for this proposal is part of a growing willingness by many people to inject the government into the activities of our private sector. It seems that everywhere we turn today, someone is calling for the government to do what our economic system has previously asked the private sector to do — whether that be government allocation of credit or government control of price and supply of resources or government redistribution of wealth or government determination that the oil industry should be broken up.

What I am saying is that I believe the explanation for the divestiture movement runs deeper than a politically motivated attack on the oil industry. It reflects a more serious and fundamental problem affecting our society today, namely a serious weakening of the confidence of many Americans in private enterprise as the most efficient vehicle for furthering our economic objectives.

We seem tempted to turn more and more to the Government to solve our real or perceived problems at whatever cost — either to our pocketbooks or more importantly, to our freedom. Not enough people have recognized that more often than not, government "solutions" lead to further problems and yet more government involvement can undo the effects of earlier "solutions." Our 200 years of experience with the free enterprise system in the United States has demonstrated that the system works, and works better than any other economic system in effect anywhere in the world. It feeds, clothes and houses more people more efficiently than any other. Most important of all, our free enterprise system also serves as the underpinning of our free society. The fact is that in every country in which the government's share of economic activity has increased there has been a tendency to move toward instability, toward minority government and toward a threat to the continuation of a free society.

The area of energy policy provides a good example of the problems caused by excessive government control and intervention. The simple fact is that this country has been without a coherent energy policy for too long — not because we didn't know what to do from an economic standpoint but because we have lacked the collective political will to take the necessary action. Make no mistake about it, policy in the energy field cannot be made by the executive branch alone. It takes the cooperation and support of the executive branch, the Congress and the American people. In January 1975, the President put forward a comprehensive energy policy. It called for maximum reliance on market forces in order to increase domestic supplies, reduce wasteful use of energy and minimize our vulnerability to foreign suppliers. Unfortunately, it was not accepted and much of the energy legislation that has been proposed or enacted would move us toward more government control rather than less.

Again and again, we seem to choose the easy solution — to legislate new controls and more government regulations. However, the experience of the past two or three years should demonstrate that further controls will only handicap and impede the energy industry in its ability to increase energy supplies and reduce our reliance on expensive and unreliable oil imports.

Similarly, the politically easy solution is to seek to maintain energy prices below their competitive market levels even though this encourages continuation of the wasteful use of energy and discourages development of new supplies. As is often the case, the appealing solution politically is not always the right one economically. This brings me back to the main subject of my talk, divestiture. In my view, the divestiture proposals now before Congress reflect a "solution" which may be politically attractive, but is counterproductive to our energy objectives.

General Arguments For and Against Divestiture

The basic arguments of the proponents of energy industry divestiture are couched in terms of enhancing competition. For example, the preamble to one of the recent bills argues that "existing antitrust laws have been inadequate to maintain and restore effective competition in the petroleum industry." So it is proposed that the laws be changed "to require the most expeditious and equitable separation and divestment of assets and interest of vertically integrated major petroleum companies." Another bill is designed to "create competition in the petroleum industry, thereby, breaking the economic stranglehold of monopoly power" and "to prevent in advance the aggrandizement of monopoly power over alternative domestic sources of energy."

The Treasury's strong support for strengthening our free enterprise system should leave no doubt about our desire to maintain and enhance competition in our economy. In my judgment, however, our antitrust laws, which are designed to ensure that competition is fostered on an industry-by-industry basis, have been effective, and I support their continued rigorous enforcement. However, the preambles of most of the recently introduced energy divestiture bills imply that our antitrust agencies have been dilatory and ineffective because they have not found sufficient evidence of monopoly power in the oil industry to support a national antitrust complaint under existing law — so Congress needs to take independent action. In effect, the Congress is legislating a guilty verdict and a harsh penalty without trial.

It seems to me that the positions of the antitrust agencies can also be taken as evidence that effective competition does exist in the oil industry. The type of incursion into our economic freedom reflected in these divestiture proposals would be quite serious in any event, but to pursue such a disruptive course in the critical energy industry at a time when our economic and our national security require a strong domestic energy industry able to reduce our reliance on foreign oil makes bad policy even worse.

Conclusions of the Treasury Analysis

The central question in the divestiture debate is whether any benefits that might be derived from divestiture outweigh the anticipated considerable costs of such a measure. The proponents of divestiture claim that it will increase competition, leading to lower energy prices, greater energy supplies and a reduced influence and dominance of the oil producing countries. We have concluded quite to the contrary: that with divestiture, it is more likely that domestic prices will increase instead of decrease and that domestic energy supplies will decline rather than rise. In addition, we have concluded that divestiture, particularly vertical divestiture, would seriously affect the ability of a major sector of our energy industry to meet our energy supply requirements, particularly over the critical period of the next 10 to 15 years. The result, we have concluded, is that divestiture will increase our reliance on imported oil and that OPEC influence over the international energy market will likely increase. We, therefore, have come to the conclusion that divestiture would hinder the achievement of our national energy goals and would not be in the public interest.

Detailed Findings of the Treasury Analysis

Let me now outline a few of the specific conclusions of the Treasury divestiture analysis. As I noted, we have examined the financial, economic, legal, and energy market effects of the vertical and horizontal divestiture proposals. We considered the likely effects both in the transitional period, which could be quite lengthy, and in the longer run when a new equilibrium will presumably be reached. The financial effects would be most pronounced during the transition period but there would be many adverse financial effects in the longer term as well.

Transitional Effects - Legal. In examining the consequences of divestiture during the transition period, we looked first at the legal and administrative problems in implementing divestiture under the legislation before Congress. Divestiture, of course, has been used as an anti-trust remedy in the past and the resulting legal and administrative problems, while complex, have been manageable. What is different in this case is the scope of the undertaking, the nature and structure of the affected industry, and the critical time at which divestiture of this vital industry would be ordered.

Although the leading divestiture bills call for a transition period of 5 years for vertical divestiture and 3 years for horizontal divestiture, legal challenges to the constitutionality of the legislation and to the fairness of specific divestiture plans could suspend or impede full implementation of divestiture until due process is given and the legal issues resolved. Thus, it is likely that the transition period could extend for 10 or more years.

Another issue that likely will be the subject of extensive and lengthy litigation is whether existing loan covenants and indenture agreements are actually violated by divestiture plans. Lenders who are relying on a company's overall credit-worthiness as security for investments may see their interests as being adversely affected under divestiture and might litigate or attempt to enforce their rights under existing loan agreements which generally place restrictions on the sale or spin-off of assets. While negotiated solutions to such problems with lenders will eventually be arranged in most cases, creditors would be in a position to threaten to accelerate the repayment of outstanding debt. In some cases, they may decide acceleration is necessary to protect their interests. In addition, the negotiated solutions which are achieved will likely often entail shorter repayment schedules, security against some of the corporation assets and higher interest rates. There are a number of other difficult legal problems which are important to consider, including particularly difficult problems relating to foreign entities and the treatment of the foreign assets and liabilities of U.S. companies.

Transitional Effects - Financial. Eventually all of these problems will be resolved, but it is important to emphasize that there will be a lengthy period of uncertainty about the structure of the new firms, their relationships with existing creditors and equity owners, and their future creditworthiness, all of which will have a detrimental effect on the access to capital markets and availability of external capital to these firms. Moreover, the capital that is available both during the transition period and afterward can be expected to cost more, due to the increased risk and uncertainty as to the future prospects of these firms.

Specifically, we believe that the financial effects of divestiture upon the affected companies during the transition period would include the following:

First, it will be difficult for new unsecured long-term debt issues (including the refinancing of maturing issues) to be sold until lenders could ascertain: (1) what corporate entity would be responsible for debt repayment, and (2) what the existing assets and liabilities of that corporate entity would be. Under current bills, this hiatus could run 1-1½ years, or longer if legal delays are encountered.

Second, even after this point is reached, it still may be difficult for some companies to raise capital since uncertainties will prevail also about the earnings potential of the divested companies. The return to divested oil companies might well be lower since economies of scale which have been advantageous to the integrated oil companies may no longer be available. Efficiencies which were possible in a larger, diversified company, for example, in such activities as planning, resource allocation, and research may no longer be available.

Third, it may be possible for these companies to raise some amount of secured long-term debt. However, since the basic security of loans would be the particular asset rather than the creditworthiness of the company, the potential volume of such financing would appear to be limited by the specialized nature of many of the oil companies' assets and possible efforts of creditors to block such financing in order to protect their existing investments.

The financial problems will clearly vary from firm to firm, but generally, we believe that there will be a significant reduction in the ability of these firms to finance energy investments during this period. Moreover, and just as important, corporate management will have to direct a significant amount of its effort and attention to preserving or realizing values of assets rather than expanding energy supplies. As a result, priorities for the vigorous expansion of domestic oil and gas resources will be downgraded, which will delay the development of these resources and result in a continued increase in oil imports. While we would expect that most projects now underway would be completed, the uncertainty created by divestiture could delay, and perhaps prevent, some projects. This would be especially true in the case of projects where financing was not completed. For example, divestiture might create major problems for the timely completion and financing of the Alaskan oil pipeline.

My discussion thus far has concentrated on the financial affects of *vertical* divestiture, for it is this form of divestiture which would have by far the greater effect on the ability of and incentive for affected companies to make investments. *Horizontal* divestiture, mandating the divestment of non-petroleum energy sector operations, would also adversely effect the levels of investment in the non-oil energy operations of these companies (i.e., the operations that should be developing the important alternative energy sources this country requires).

Long-Term Effects. It is the transition period effects on investment that we see as the most critical. This transition period, which, I would emphasize could extend for the next 10 years or more, is the same period during which the domestic energy industry must make massive investments if the nation is to reduce its dependence on foreign oil. In our judgment, the adverse transitional effects are a sufficient reason for opposing divestiture. However, we also examined longer-term, post-transitional period financial effects of divestiture. *First,* the existing integrated companies would have a greater debt capacity than the aggregate debt capacity of the divested component companies since an integrated company has greater stability in its level of cash flow and is viewed as offering a greater likelihood for principal and interest payments on debt to be met. Also, in the case of horizontal divestiture, the divested non-oil firms would lack the financial backing of their former parent firms. *Second,* in the case of vertical divestiture, the required levels of working capital probably would also rise. *Finally,* with respect to incentives to invest, the size and output thresholds imposed by divestiture would effectively place growth ceilings on firms approaching those limits.

Effect of Divestiture on Energy Objectives

These are some of the financial aspects of divestiture which we believe should be given careful consideration. As I mentioned earlier, however, the ultimate test of whether we should undertake divestiture is whether it would further our energy objectives of increasing supply at the lowest cost to the American consumer. In this regard, I believe the burden of proof should be on those who are calling for this costly restructuring of the energy industry to establish the benefits that would result. I have seen no such evidence. I think it is time that we begin to ask some of the right questions:

First, would divestiture result in an increase in competition in the oil industry? Clearly, it would increase the number of firms in the industry and end the corporate ties between the functional components. But what has not been shown is how this will result in lower domestic energy prices or increased domestic supplies. The horizontal structure of the oil industry will not necessarily be changed by vertical divestiture. Thus, even if there were an abuse of market power in one *horizontal* level which existing antitrust laws or, regulation could not handle, *vertical* divestiture would certainly not be the appropriate remedy. Moreover, if our antitrust laws or our regulatory policies are deficient in any way, then such deficiencies should be rectified. We should all support the rigorous enforcement of our antitrust laws and sound regulatory policies but let's not substitute divestiture legislation for such a policy.

Second, would divestiture result in lower domestic prices through increased competition? We have been unable to uncover any evidence that divestiture would lower prices. On the contrary, our analysis shows that the likely direction for oil prices as a result of divestiture will be up, not down. The likely adverse effects on investment in new energy supply capacity would tend to put upward pressure on prices directly, and indirectly, by increasing our dependence on foreign oil. Further, we have seen no evidence that there are significant inefficiencies existing in the present oil industry due to any lack of competition. On the other hand, there *are* important economic efficiencies in integrated oil operations which are recognized worldwide. In fact, because of these efficiencies, many other governments are seeking to increase the degree of integration of their oil industries. These efficiencies would be lost under vertical divestiture.

Finally, would divestiture help increase the development of alternative supply sources? Again we believe the legislation would have the contrary effect, with the financial uncertainties resulting from divestiture increasing the cost of capital to affected firms and reducing their ability to raise external capital for investment in alternative energy supply sources.

There are many other implications of the divestiture proposals which I haven't tried to cover today. We are certainly not able to answer all the difficult questions relating to the divestiture issue. At the minimum, however, we would hope our analysis will help focus attention on the very serious questions that must be addressed before such drastic action is undertaken.

Today, as never before, energy policy has become intertwined with national and international political concerns. For some, the emotions of the political arena have distorted the economic realities of the marketplace. As we consider the divestiture issue, I would urge that we minimize the superficial political rhetoric and maximize the objective economic analysis, and I believe that your conference will be of tremendous help in this effort.

5

THE ACADEMIC VIEW: THE COSTS AND BENEFITS OF DIVESTITURE

THE ECONOMICS AND POLITICS OF DIVESTITURE

Edward W. Erickson
Professor
North Carolina
State University

Thank you. When I look at the divestiture, dismemberment — divorcement — whatever you want to call it — debate, I look at it as an academic scholar trying to understand the social process in the words of the urban rock blues — "What's Goin' On." I try to figure out what is in fact going on. As I see it, before one can discuss the effects of divestiture, one has to try and understand that. As I look at things, I can see a number of potential reasons for consideration of divestiture proposals. One of them is the kind of situation that as a professional I am most comfortable analyzing, and that is "what are the economic costs and benefits for divestiture, particularly on the grounds of increases in competition." The second reason is a concern over social and political power on the part of the oil industry in aggregate, and the firms in the oil industry in particular. A third reason, which will be touched on later and which I won't dwell on except in passing, is the belief by some — I believe mistakenly — that divestiture would result in a lower world oil price. I'll leave that to Tom Stauffer and others to discuss in detail later this afternoon. And lastly, one reason which I think tends to be overlooked, at least in discussion, although it's not overlooked in terms of the way people are motivated, is a fundamental antipathy towards the petroleum industry.

Now, I wanted to address these three issues: an aspect of economics in terms of cost and benefits in competition, social and political concerns, and antipathy towards the industry. One of the important things that was not said this morning in regard to competition is that the majors compete with each other very vigorously. Senator Haskell on several occasions — in talking about oil and coal and uranium — talked about everything being all in the same hands, as though Exxon, Texaco, Gulf, Amoco, Standard of California, Phillips, etc., were all the same company and made joint decisions. They are not and they do not; they are vigorous competitors with each other.

Another point that came up this morning, is the idea that competitive pressure in the industry for decreases in the price of gasoline, for example, stem from the independent sector. I regard that as at best partially true and at worst, conceptually erroneous. It is true that there was a substantial decrease in the real price of gasoline over the last quarter century. From 1950 to the early 1970's the real price of gasoline fell by over 25%. Now, the real price of gasoline fell because supply expanded relative to demand; and there was substantial competition among the majors in terms of their moves to expand, contract, and adjust to changing market conditions — all on the premise of trying to make a buck. And I think that is desirable. If one tries to examine the proposition that the competition stems from the independents, the logical fallacy is that at the margin where prices are determined one cannot identify an incremental barrel. All barrels or all gallons are equally responsible at the margin. To make a bad paraphrase of Ernest Hemingway's own borrowing from John Donne, the question is — "ask not which is the marginal barrel, because all barrels are equally responsible at the margin." And in that case the majors have participated in the expansion of the industry; if one then ignores my advice and wants to attribute responsibility for the increases in supply that have put the downward pressure on prices that we saw before the OPEC revolution, one has to attribute two-thirds, three-quarters, of the responsibility to output expansions on the part of the major oil companies.

Another aspect of competition that I want to touch on is the often alleged practice of cross-subsidization. My co-speaker, Professor Patterson, has written alone and with Professor Allvine on that. I will quote first from Professor Allvine. He said, "Very often the integrated oil companies earn little or nothing on the downstream investments while crude oil is very profitable." Switching to Professor Allvine's alter-ego, Professor Patterson, "Profits were lodged in tax sheltered crude oil production, and for one's own crude to flow it was necessary to gain control of over downstream markets and to maximize the flow through one's own outlets." At the same time Professors Allvine and Patterson correctly observe a 2 to 5¢ differential in gasoline prices between branded sales and non-branded sales by independent marketers. Now, when you buy gasoline you are buying not only gasoline. There is a distinction between service stations and filling stations and it is possible to buy a whole bundle of attributes along with a gallon of gasoline. It seems to me we have evidence that a logical interpretation of the data before us would suggest that what consumers were buying was what they wanted to buy. The consumer sovereignty was at work and the companies were delivering a bundle of attributes with gasoline which was what was desired. In that sense, rather than there being inefficiency in marketing, there was efficiency in the sense of consumers' wants and desires being met. Why is that? Well, look at this statement that "in order to gain control over downstream markets and to maximize the flow through one's outlet, you have to engage in unprofitable marketing in order to have an outlet for crude oil."

With a 2 to 5¢ differential between branded sales and non-branded sales, if consumers wanted something other than what the branded marketers were selling, we would expect them to respond to that differential. If they responded to that differential, then the market share of the non-branded marketers would have expanded and one would not have had to "have control over your own downstream outlets" to push your crude because the independent marketers, expanding as a result of the 2 to 5¢ price differential and consumer response to that, would have provided that outlet. At the same time we would then see two benefits to the major oil companies under the kind of scenario that some people seem to believe exist. First, because of the downward slope of the total demand curve for gasoline more gasoline would have been sold; that would have meant more crude oil would have had to be produced, and companies would have made more profits on their "profitable crude oil production." And second, the companies would have been able to do without the "unprofitable marketing operations" that they are alleged to have engaged in. The upshot is that I have a logical problem with the idea of cross-subsidization.

I also have a factual problem with the data that are often used to support the idea that one segment of the industry is subsidizing another. That is something we may discuss in the question period. I just don't think the data really show that.

Now then, with regard to social and economic power, to say that the industry is competitive is not to say that we have necessarily had a Pareto efficient allocation of resources. One has to talk about competition at a given level of cost; there have been substantial regulations which have affected the levels of cost at which competition has taken place. Those regulations, for example, include depletion when it existed, and the associated tax on benefits, the mandatory oil import control program, the market demand prorationing activities of the Texas Railroad Commission, and the activities of the Federal Power Commission to hold the price of natural gas down. There is a demand for a regulation. The fact that there has been regulation which has been "good for the industry," is evidenced by the circumstances that when Lyndon Johnson was the majority leader in the Senate, and Sam Rayburn was speaker in the House, that the industry did have a lot of clout. But the questions are "what were the costs and benefits?" and "what was the distribution of the effects of regulation within the industry in terms of who bore the costs and benefits of that regulation?" What I would submit is that in general — and I must admit that this is a generalization subject to all of the fallacies of any generalization — the costs of regulation were borne by the major oil companies and the benefits of the regulations that were spawned by the social political power of the industry accrued to the independents. Independent producers, independent refiners and marketers. For example, under market demand prorationing, the large efficient production characterized by the major oil companies was shut-in to make room in the marketplace for the stripper production of independent operators. The mandatory oil import control program had the sliding scale provision in it to the benefit of the smaller refiners. We have the defense supply purchases set-aside provisions which always favor small business. In the Federal Power Commission, in the administration of the regulation of well-head price of natural gas, there are various administrative procedures which favor smaller producers over larger. The intent, although it would have not been the effect of the natural gas bill that the House passed on February 5 of this year, clearly was to make a distinction between large and small producers in favor of small producers. Look at the mandatory oil import control program. Whose oil was being kept out of the United States? It was the major oil companies at that time, producing in Venezuela, Saudi Arabia and elsewhere. If one goes through the record of public policy, and tries to track social and political power back to its origins, one does not arrive at the doorsteps of Exxon, Gulf, or Texaco, at all; one arrives at the doorsteps of Texas Independent Producers and Royalty Owners Association, and other independent agencies.

My feeling about that is that if we were to divest the industry, or to divest the major firms, the industry would still exist. The political process would still be what it is. If anything, there would be an increase in the probability of public policies, which would cause further differences between Pareto efficiency in the allocation of resources and what an economist would actually prefer to have occur.

Finally, I'd like to deal with the problem of fundamental antipathy toward the industry. It seems to me to be in a strain of American fundamentalism that when facing a problem we typically try to solve that problem by dividing things into good guys and bad guys, with white hats and black hats. In fact there is somebody running for one of the nominations for the presidency who has a long career in that activity. Our typical response is that someone is to blame, someone must be punished; if the person to blame is not readily available to punish then we'll punish someone who is handy. Maury Adelman was referred to a number of times this morning. Maury, I think, had a very appropriate comment on this fundamentalist punishment syndrome in the hearings before the Multinational Committee when he said the vituperation that was being heaped upon the oil industry was analogous to the proceedings during the McCarthy era.

Now, what would be the effects of divestiture and horizontal disintegration. Well I do not have an econometric or input-output model into which I could crank divestiture and crank out a new set of outputs which would show employment effects and costs and that sort of thing. There would be costs. They are very difficult to specify in precise quantitative terms. I can see absolutely no benefits whatsoever on any economic grounds. On that basis, the benefit-cost ratio clearly is unfavorable. On a more conceptual basis, divestiture which would prohibit firms from being integrated in the various stages of the petroleum industry or horizontal disintegration, which would prohibit oil companies from aggressively expanding supplies in coal (and I found Bill Slick's comments this morning very persuasive), would be a restriction on the mobility of resources. A restriction on the mobility of resources cannot have good effects. It can only have bad effects, leading to a resource allocation efficiency less than what it otherwise would be. Moreover, I think that one of the critical problems with divestiture would be the uncertainties it would entail and the fact that it would delay the development of alternative energy sources in the United States through expansion of conventional oil and gas production, adoption of enhanced recovery projects and other activities that the companies may choose to engage in. The effect of that can only be to increase the probability that OPEC's price discretion will be larger than it otherwise would be, and that the world price of oil will be higher than it would otherwise be, rather than lower.

Now, I think all of those are important adverse effects of divestiture and disintegration; but they look at divestiture, divorcement, disintegration, dismemberment — whatever you want to call it — primarily as an economic phenomenon. I don't really believe that the reason we are all in this room today and the Congress is considering the bills is primarily an economic phenomenon. To answer my own question of what's going on, one falls back to the American fundamentalism that I referred to earlier. What's going on is a move to punish big oil. Divestiture would certainly do that, but in my opinion, that would be like cutting off our nose to spite our face. Thank you.

THE ADVANTAGES OF CRUDE OIL DIVESTITURE

James M. Patterson
Professor
Indiana University

After hearing the comments this morning, and after listening to Professor Erickson, it is clear that I ought to shift gears from what I was going to talk about and deal with the very specific competitive gains that follow from divestiture. I teach marketing, so the part of the elephant I have hold of is different from the part that Professor Erickson has. Perhaps that helps explain our differences.

The forces of competition that characterized this industry in the 1960's largely arose from the surplus of gasoline that was generated by the major refiners. The gasoline that the gadfly independent, or unbranded marketer sells typically comes through a crossover process from the majors. The major-majors sell their surplus to the minor-majors who in turn sell their surplus to the independents. My concern is that this situation has changed. The repeal of the depletion allowance and the emergence of the three-tiered price structure for crude has now radically changed the incentive for the refining sector to produce surplus gasoline. This loss of incentive represents a fundamental weakening of the forces for competition at the marketing level.

To show what is happening as a result of these changes, I jotted down some figures from the *Oil & Gas Journal*. On the first of March, the stock of gasoline was 244 million barrels; demand was 6.3 million barrels; the industry was producing at a rate of 6.5 million barrels. Now, as we approach the seasonal peak demand period, the latest figures that I have show that the stock has fallen from 244 million to 218 million barrels. The demand has risen from 6.4 million barrels to 7.3 million. But the amount produced remains essentially unchanged — 6.53 vs. 6.54 million barrels per day.

What we see is a situation where, in the face of predictably increasing demand, the industry is not responding by producing more gasoline. Rather it is drawing down on its inventories quite heavily so that in last Friday's *Oil Daily* the headline reads "Spot Gasoline Price Hits New Level." The price of spot gasoline of the Gulf in cargo lots has risen by 8¢ in the last three weeks. I think that this shows that something in the system has changed, and that the incentives to increase production have been blunted.

We have a situation in which the marginal cost of additional production is higher than the average cost. The average cost reflects the mix of the three prices — imports, new domestic and old domestic. But, if you are going to increase output, you are going to have to use imports at $13 plus. Under such circumstances, additional output beyond that needed for the integrated refiner's own channels and his 1972 allocated customers is just not profitable.

How have the independents fared during this recent period? Starting in the summer of 1973 and then certainly through the embargo of the winter of 1973 and early 1974, the independents were denied product by the normal process of suppliers funneling as much of the scarce gasoline as possible through their own integrated outlets. It wasn't until the Congress stepped in with mandatory allocations that the independents were restored to their 1972 position. Their present good growth and prosperity is both unnatural and short-lived. It is due completely to the government's interference with the workings of the industry.

At one point last year, the spread between the low-ball private brand price and the high-ball major price in Los Angeles was 11¢. That is an unusual situation. The difference in gasoline prices between Buffalo, Baltimore and the other 55 cities listed in *Platt's Oilgram* also reflect unnatural differences. This is due to the distortions that come from regulation. But there are some more basic forces at work here.

During the period of the late sixties and into the present, we find a shift in the character of vertical integration in the industry. The shift comes from the increased balance that is taking place. Beginning with the writings of J. J. Spengler and Fritz Machlup, there is a well-developed literature in economics that says that vertical integration is not necessarily bad, and that in fact it may often be viewed in a positive light since it eliminates the repeated marginalization of the revenue curve when the manufacturer sells to a wholesaler who sells to a retailer, etc.

However, the fact is that before the mid-sixties many integrated oil companies were out of balance. Many had marketing capabilities substantially less than their refining capabilities. Others produced much more crude than could be used by their refining and marketing operations. The result was that, while they served themselves, they also served others. I am struck by the fact that the testimony by Mr. Slick before the Congress last year shows that in 1973 the four largest firms produced 31% of the crude, refined 32% of the product, and marketed 30% of the refined product through their own channels. They are very closely in balance. The figures for the top eight are 50%, 57% and 52% respectively. They are a little less in balance, but not much. The result of all this is that it is becoming harder and harder for the non-integrated or partially integrated marketer, who depends on his competitors for supply, to get product without government pressure. The same thing is true of the independent refiners. They have increasingly sought to combine the profits of refining and marketing by increasing their direct participation in retailing. Again, as their balance becomes more complete, it is going to be more and more difficult for the non-integrated marketer to find gasoline — gasoline which was previously available at very good prices which were often as much as 7 to 8 cents under tankwagon.

The point that I want to make here is that there has been a dramatic shift in the character of vertical integration in this industry which doesn't show up in the traditional concentration measures. The result of this is that the private-brand gadfly, who has been the instrument for regulating marketing margins in the industry, is now in the position where, were he not protected by allocation programs, he would more and more be in a position where he would not have sufficient supply to provide him the incentive to cut price in order to increase volume. This is the damage to competition that I think is critical.

Many say you are only talking about a couple of pennies, but with more than a hundred billion gallons of gasoline sold each year, the pennies mount up. Compare this with the figures that Professor Erickson and others have been producing to show the costs of divestiture.

The possibility of saving several billion dollars from increased retail competition is not impossible since there is clear evidence now that some of the mass marketers can operate profitably on a rack to pump margin of 4 to 5 cents. If you look at the major brand marketing margin, including wholesale and retail margin plus the refiner's marketing margin, it is on the order of 13 to 14 cents. This is a potential gain to the consumer of 8 cents per gallon if the market were freely allowed to adjust.

While some of the price differentials reflect the differential service that is being offered, the fact remains that were it possible for the gadfly to have unlimited access to gasoline, the spread would be dramatically eroded. But given the present integrated structure of the industry, and as soon as the allocation controls come off, there is going to be a limit on the amount of gasoline available to the price-cutters. This in turn is going to reduce their incentive to take less margin and to drive for higher volume. This is why we need to change the structure of the industry. A simple change is all that is needed. It seems to me that the separation of crude oil production from refining and marketing would make it possible for new refiners to enter the market without having to integrate back into crude and that this would tend to dramatically increase the supply of uncommitted gasoline available. An increase in the supply of gasoline in turn will increase the incentive for the gadfly to cut price and attack the prevailing margin.

As long as the control over the price of the incremental supply of crude is held by OPEC, the price of all infra-marginal crude will rise to that level. I don't think that preserving or changing the structure of the domestic industry is going to change that. I do think a change in structure will change the marketing margin and that that would be a worthwhile benefit.

In terms of political strategy, it seems to me that the oil industry might want to consider crude oil divestiture. So long as uncommitted gasoline is going to be in short supply, there is going to be a fight. Because of vertical integration, the market solution will not be seen as satisfactory. Consequently, the allocation will be made by the government. With crude divestiture, the decision could be made by the market.

I believe there would be a very positive gain to the industry to get the government out of their business. Crude divestiture would facilitate this. So long as most of the crude oil and most of the refined product is kept off the open market and directed through integrated channels, there is going to be continual public concern over how the decision about how much to route through integrated channels and how much to let get on the open market are made.

DISCUSSION

William A. Johnson, Professor, George Washington University

The two speakers raised several important points that arise from the possible vertical and horizontal divestiture. Let me zero in on several statements that I would like to magnify or correct, and then add others that I regard as important likely consequences of vertical and horizontal divestiture.

First of all, the speakers did focus on the issues of competition, on whether vertical divestiture will make the industry more competitive than it is at the present time. Second, they talked a great deal about the allegation, which has been with us for years, that there has been subsidization between various vertical segments of industry activities. In particular, they suggested that profits made out of production have been used over time to subsidize downstream activities and, in this way, to provide a competitive advantage to the integrated companies' outlets relative to independent marketers.

Subsidization of marketing

The cross-subsidization argument has been widely discussed by a number of people. The basis for this argument has been undermined in recent years with the dropping of the depletion allowance, termination of the oil import program, and other policy changes. Perhaps more to the point, the argument was never correct, a fact that has been well demonstrated by Richard Mancke of Tufts University. Mancke shows that in order for the integrated companies to benefit by subsidizing downstream activities over the long run, they would have to be substantially self-sufficient in crude oil production, somewhere in the vicinity of 94%. Among the 18 companies that would be affected by vertical divestiture legislation, only one is more than 94% self-sufficient. The average of the integrated companies is about 60%. These companies have no incentive whatsoever to subsidize downstream activities by maintaining high profits at the production levels.

However, I think Professor Patterson is right when he states that price controls have generally caused marketing and especially refining activities to be uneconomic in recent years. The most important culprit, in my judgment, has been the non-product cost passthrough regulations and several other regulations by the Federal Energy Administration. But is Professor Patterson correct when he asserts that if we removed all the controls, we would then see the integrated oil companies putting the independent marketers out of business? I don't think that is the case. At least three of the integrated oil companies that I know of at the present time would like nothing better than to get out of marketing if they could just get rid of the govern-

FEA regulations

ment's allocation program.

Independent marketers

In all three cases, the companies cannot compete against the independent marketers. Most independent marketers have an advantage in that they know and understand local marketing conditions. Corporate bureaucrats from the large integrated oil companies simply do not. There is a role for the jobber and a role for the local dealer. Although there are major changes taking place today in marketing, I doubt very much whether the independent jobber or dealer will disappear. My own guess is that economies that accrue to the independents, and an understanding of the local market, will be sufficient to keep them competitive with respect to major oil companies.

Moreover, if controls are the major reason the oil companies are not producing surplus gasoline at the present time, then I do not see why the problem should continue if you get rid of controls. I strongly suspect that the economies of scale in refining operations might reassert themselves, and after a certain period of time we would, in fact, see surplus capacity in refining.

Let me now make several other observations about the possible economic impact of vertical and horizontal divestiture. Professor Erickson made a point which I think is very important and should be underlined. The major impact of this legislation

Uncertainty after divestiture

is uncertainty. Uncertainty is a dead weight upon capital markets, and it is undoubtedly going to discourage new investment at a time when we vitally need it. The reasons, I hope, are obvious. We are going to have at least a decade of litigation emerging from divestiture legislation, notwithstanding the creation of special courts, the delegation of substantial powers to the Federal Trade Commission or some other government body to clear away the disputes that are created.

Let me give one example concerning debentures and bondholders. The divestiture of the integrated companies is going to

Capital formation

result in the destruction of a major asset of bondholders. Virtually all debenture agreements contain covenants or are backed by understandings which restrict the rights of the integrated oil companies to dispose of major assets without bondholder approval. Divestiture would force a disposal of assets. The bondholders would have the right, unless it is taken away from them by the government, to accelerate the repayment of debt. For the 18 companies, 15 to 20% of assets are covered by debentures. To cover these debentures, the oil companies would have to liquidate assets. And, if they did not liquidate assets, some could be forced into bankruptcy.

The result of this will be the forced diversion of liquid assets away from the oil industry. The companies will have to focus on paying their bondholders and every penny that the companies earn will have to go into bond repayments as opposed to investing in new productive capacity. Given our experiences of the last several years, the industry's capital should be going into investment, not liquidation.

My final point is that divestiture will have a deleterious effect on energy research and development. Typically, research and development taking place in one segment of the industry benefits another segment. Research and development expenditures are also a function of firm size. Divestiture will have a deleterious effect, particularly on the development of long-term alternative sources of energy such as coal gasification and liquefaction.

Energy research and development

John Wallach, State Department Correspondent, Hearst Publication

I'd just like to ask Professor Patterson one simple question that has been raised several times today, but hasn't been adequately answered by the advocates of divestiture. Why should oil companies invest over the next five or ten years in expensive high-risk exploration of oil, and additionally in other potential resources, to achieve a greater self-sufficiency than we have today if they are faced with the Justice Department and time-consuming litigation? What incentive will prompt the oil companies to take a questionable and expensive risk in trying to achieve greater self-sufficiency if they are going to be consumed with another expensive operation — that of defending themselves in the courts against divestiture?

Investment uncertainty

James M. Patterson, Professor, Indiana University

I am not sure I am the person that can answer that question, but I think it assumes that these divorced segments will be unprofitable and I think that's unfounded. I think that the divested crude producing companies will perhaps be even more profitable. In fact a study done by one of the investment

Profitability of divested sections of oil companies

advisory services showed that if you took the assets owned by the integrated companies and value them at the same rate that the stock market values the assets of a non-integrated company — by number of substantial wells or barrels per day of refining capacity — the integrated company would probably increase in stock market value if dismembered. I am not *sure* that's true. On the other hand, I don't think it's reasonable to assume that there will be less profits.

John Wallach

I think your response begs the question because the oil companies will be faced with the prospect of losing any potential profit during a period of perhaps 10 or 20 years. Why then should they take the risk?

James M. Patterson

I don't think they have the right to profits that come from market power. They have a right to profits coming from serving consumers' interest and from efficiency, but they don't have any inalienable right to preserve market power profits.

William A. Johnson

Investment after divestiture

We should not only be concerned with the oil companies, but also with outside investors. I've seen estimates that as much as 40% of the capital that's going to be needed by the oil industry is going to have to come from external sources. Will bondholders be willing to lend money to oil companies through this long transition period during litigation, when there will be continued uncertainty, and during which, I strongly suspect, the various covenants that are written into debentures will be destroyed in order to protect the oil companies from a run on their liquidity? Will shareholders be willing to invest during this period? I seriously doubt it.

Edward W. Erickson, Professor, North Carolina State University

Market power

Since we are talking about implicit assumptions, I'd like to approach one of the explicit assumptions that Professor Patterson raised with regard to power. In my opinion there are entirely too many oil companies — they behave too independently for there to be any market power. If oil companies have market power they are in fact the dumbest exercisers of market power that I have ever seen in my life. It is a well-known proposition, one we teach to sophomores, that somebody with market power prices in the elastic portion of the demand curve; and the demand curve for gasoline — nobody knows exactly what it is — is probably negative .2. So those guys have got to be stupid.

Stephen H. Goodman, Vice President, Policy Analysis, Export-Import Bank

The only clear point that has emerged from the discussion so far is that the greatest argument against divestiture is its administrative and financial difficulty. Clearly divestiture would impose some real costs. I think therefore it is the obligation of those arguing for divestiture to demonstrate that there are benefits other than for the lawyers who will be involved in litigation. We always hear about the benefits of increased competition and of course competition is a bit like apple pie. It has a great deal of political appeal, it has sociological appeal and, as one of our speakers said this morning, it is part of our national policy. Frankly I think the kind of influence that the oil companies may have is somewhat different than what most people tend to assume. I don't think the collusion or market power, to the extent that there is any, is aimed at fixing a price per se. I think it is more aimed at a control of the environment. They wish obviously to gain more control over the various states of production; they wish, to use Mr. Patterson's terminology, to balance their integration in order to have more control over their environment.

Administrative problems with divestiture

I don't think we have any evidence that greater control over their environment results in either higher or lower prices. I think we need more study. As an economist, I guess that's self serving — instead of having the lawyers employed for the next ten years, let's have economists.

I think, though, with that in the background you may be drawing too sharp a distinction here. We are only offering two options. One is divestiture, which will have many costs in the short run, and possible benefits — who knows? — in the long run. The other is to leave things as they are now.

A third option: regulation

I'd like to propose a third option between these two extremes. I think all of the speakers — Mr. Parsky made this point very clearly at lunch — make a distinction between vertical divestiture and horizontal divestiture. Horizontal divestiture would, it appears, involve much less cost to the economy. I think that's one distinction we should make. Second, let's make a distinction between divestiture and regulation. I think divestiture involves substantial financial costs; regulation may not involve these costs. By regulation I am thinking of limiting the acquisition of new assets by the petroleum industry in non-oil energy production. I don't think many of the costs attributed to divestiture would occur under this regulatory solution. What advantages does regulation have? It does appeal to people's political sense. It also prevents the threat of market power. Senator Haskell this morning expressed a general and widespread concern that if there was a surplus of energy from a number of sources the producers of energy would, to the extent that they could, seek to restrict supply. This proposal would attack that concern without imposing some of the costs.

Regulation of horizontal investment

I think there is also an economic and a free enterprise argument in favor of regulating oil company acquisition of non-petroleum energy resources. The argument is that an industry has certain specialties, and capabilities in the market in which it operates. We shouldn't tear asunder what the market forces have built up, but I don't think that same idea translates into the maxim that an industry should use its cash flow as it sees fit. Most people would say that if an industry can't profitably invest its cash flow in its own activities, it should return its cash flow to its investors who then, through the market mechanism, make a decision about where these new investments should be put. I think regulations which restrict oil companies from acquiring assets of other energy sources would either put these companies into the position of investing more in the oil industry, drawing less capital from capital markets, or paying higher dividends. The last two options would make more capital available for new enterprises or expansion of the enterprises in these other energy areas.

Mr. Erickson, I guess I'm somewhat disappointed you didn't have any conclusions on the impact of either type of divestiture. I wonder if you would have the same cost objections to regulation, as I have outlined it, as you would to divestiture?

Edward W. Erickson

Yes. I've got another alternative which we'll call a fourth

alternative, but it is not between leaving things as they now are and vertical divestiture. I think you can substantially improve on things as they now are. The United States desperately needs substantial increments in our real energy capital. We're faced with a substantial capital formation problem. Divestiture or increased regulation will create additional uncertainties. More importantly there has been an erosion in economic incentives for capital formation. If you look at the lower tier prices, and compare them to the reference price that the Cabinet Task Force on import control used in 1969, that price was $3.30 a barrel for 35 South Louisiana Sweet crude oil at Tidewater. If you adjust that price for changes in the price level and sweep out percentage depletion, that $3.30 a barrel turns into, in 1976 dollars, $2.60. Under those situations I am not at all surprised that we've seen an erosion in the crude oil producing base in the United States. If you make the same calculation on upper tier prices, rather than being the handsome numbers that people talk about, you're talking about prices in the range of $5.50-$5.75, in constant 1969 dollars. We have had, in my opinion, a reign of error in energy policy. We have been living too long on the principle of our energy capital. We need substantial capital information and I think one of the ways to do that is to de-regulate completely the price of crude oil.

Real domestic crude prices have fallen

Deregulation of oil prices

The incremental cost of energy to the U.S. economy is at least the price of imported oil; the real cost is something more than that because of security considerations. Oil prices ought to go to the world market price tomorrow; they should be there today. Natural gas prices should go the same way. We need to eliminate the allocation program in crude oil. It makes abso-lutely no sense whatsoever for a refiner with a 25,000 barrel a day refinery to buy some used refinery equipment and expand his capacity to 35,000 barrels a day to get an allocation to pro-duce that additional crude when instead someone else could run that incremental 10,000 barrels in a 200,000 barrel a day refin-ery. It's patently preposterous to have policies that force us in these directions. What we need is not the third proposal of regulation that you suggested but a fourth proposal, which is to move in a direction arresting the trend towards Anglicaniza-tion of the American economy.

William A. Johnson

Benefits of
horizontal
integration

May I comment on that also? First of all I am going to steal Ed Erickson's term "reign of error," with your permission Ed. I think Mr. Goodman's proposition is a bad one, because no clear distinction exists between one source of energy and another. Horizontal divestiture implies such a distinction.

Our supplies of oil and natural gas are far more limited than our supplies of coal. One of the reasons why oil companies have been interested in coal, aside from the fact that coal is the only fossil fuel not presently regulated in the United States, is that as they look down the road they see coal as a major substitute for oil. With some technological developments, particularly in gasification and liquefaction, this could very well become a major input into the refining process. It is future vertical integration that the companies investing in coal have in mind.

A substantial portion of the research and development in new processes, particularly for coal gasification and liquefacation, has come from the oil companies. If you told the companies, "that's it — you can't go any further with coal," you might very well discourage the use of coal and the development of these processes.

Let me make one last point on coal and horizontal divestiture. The coal industry is simply not capable of producing by itself the amounts of coal that will be needed. You're talking about an industry that hit peak production in 1947. Since 1947 coal production has been trending downward except for this past year. Now all of a sudden, with Project Independence and the Arab embargo, coal is the great energy hope of the United States. Everyone is talking about developing our coal industry. It's been like giving an overdose of Geritol to an octogenarian.

In fact, if the industry is to do the job necessary to develop coal, in my judgment both independent coal companies and coal companies owned by the integrated oil companies will be needed to do all they can. The integrated companies have the capital. To draw artificial barriers between various segments in the energy industry is condemning the oil industry to investing in Montgomery Wards, to buying up Irvine land companies, and to other activities that are not going to help the United States in its energy difficulties.

A question from the audience:

Don't you see a number of changes coming at the marketing level, particularly among full-service outlets, that will have effects similar to those intended by divestiture, but which will come competitively and without legislative involvement?

James M. Patterson, Professor, Indiana University

I quite agree with you. The principal problem with the kind of station that reached its high point in the late sixties, the branch station with multiple bays, was a problem of volume. A large proportion of the costs are incurred when you open for business; they do not vary closely with the volume of business that you do. Efficiency of the mass-merchandiser has been achieved either through tying-in with other kinds of facilities such as convenience stores, or else by operating "gas and go-no service" stations in which you run very high volumes to cover those costs.

Future of gasoline retailing

I quite agree that when the new era emerges, when Federal Energy Administration stops playing games with marketing (which may be fairly soon) you are going to see a dramatic decline in the number of stations. I predict a drop down to 150,000 stations. The old dealer system assumes that the mechanism for tying all of these costs together — the credit, which costs up to 3¢ a gallon, the property, and so on — is the "tank wagon" pricing system. A number of minor majors, Continental and Cities Service for example, are proposing to break away from that and go to "rack" pricing. Sun has talked about a franchising arrangement. I have no reservations at all in predicting that there is going to be a virtual revolution, no matter what happens about divestiture, in the retailing of gasoline during the next ten years.

It is going to radically shift from a situation dominated by the Shell station of 1968, giving games and credit and stamps, to something radically different. The mix of marketing strategies is going to change. There will still be full-service stations serving a local market; there will still be jazzy stations on interstates emphasizing credit and high prices but there will be a much more diversified mix. The majors who in the past largely focused on the full-service market segment, and fought like mad to prevent the others from growing, are now going to be active participants in all those segments. I also think they are going to see a growth in direct operations, in secondary brands, and in thin markets. Instead of withdrawing from retail activities you are going to see the majors shift to secondary brands where they still have good exchange relations and where they can get gasoline competitively.

One of the more interesting things is going to be the entry into gasoline marketing by the traditional general retailers such as Sears and Montgomery Ward. Tie-in retailing is going to grow, and the traditional retailer is probably going to find it is easier to market gasoline than the oil companies are to get into the grocery or the auto parts business. I think that a natural divestment in marketing will take place.

I feel that the majors in the next ten years will find it in their own interest to withdraw substantially from the kind of participation in retail that they have had in the past. This is because the gasoline shortages have undermined the premium they could command as a result of brand image. With the rising prices the public is much more sensitive to price differences. I never paid attention to gasoline prices before; neither did my friends. Now it is a constant cocktail party conversation. The market is going to change and I doubt that Senator Bayh's bill is going to have any effect on it. That change is coming as soon as we get rid of the Federal Energy Administration.

Costs of regulation   An additional point is that I think regulation has created the worst of all possible situations at this moment. I would much rather trust Standard Oil to do a good job than I would to trust FEA to tell Standard Oil to do a good job. I think that radical distortions have come about through regulations. The oil industry has been controlled since Phase One; they have been the only industry subject to price controls through this period. In fact, the price of crude has been indirectly regulated since the mid-thirties through prorationing. We often think of the Texas Railroad Commission as an instrument for inflating prices. It was, however, also an instrument for preventing prices from rising. The stability of prices through the war period, except for the one or two jumps, was largely a function of demand prorationing; but as soon as someone tried to push the price up, the Railroad Commission would let more crude oil on the market and drive it back down. So, the oil industry, since the depression, has been a regulated industry.

I am suggesting it is likely to stay that way unless the industry is willing to move into a stance which looks more competitive to the man in the street. It seems to me crude divestiture would do that without major cost to the industry, and would not destroy the important economies between refining and marketing. That is where they exist.

Benefits of crude oil divestiture

The fact is that the oil companies don't tie in directly to the crude that they produce because of exchanges. Certainly, economies come through certainty of relationships, but Sohio achieves this certainty through Evergreen contracts with suppliers. It doesn't have to come through ownership. Others have suggested that even if you do divest, these supply contracts will reassert themselves. They claim no sound-minded refiner would invest a billion dollars into a refinery without some assured access to crude. In a market which is largely vertically integrated, he cannot have the assurance without a contract.

However, in an open free intermediate market for crude, where price determines who gets what, I think you could have access without contracts. Short of a free market, you are not going to have the entry into refining nor the diversity of opinions about whether or not to produce crude that independent actors would create. I am not arguing to separate refining and marketing. I am only saying separate the crude and the rest will take care of itself. I think there will be a natural divestment of retailing from refining as a result of the retailing revolution that is going to come about.

Edward W. Erickson

I want to agree completely with Professor Patterson on regulation. I also would like to agree with him somewhat on marketing. I think that in the seventies and eighties, the American consumer, will get what he wants in terms of the bundle of attributes associated with gasoline, just as he got in the fifties and sixties. I would like to qualify Professor Patterson on the question of whether or not the majors will continue to participate in marketing. I think there will be money to be made in marketing. They are in the business to make money, and we will see them do it. They may do it in different and new ways

Majors in marketing

than they have done it in the past, but that will be because they are trying to maximize.

With regard to open markets for crude oil I disagree. I think we now have open markets for crude oil. The refinery that is under construction above New Orleans, ECOL, received financing for a 200,000 barrel-a-day operation without any insured crude oil supply when that financing was arranged.

William A. Johnson

Mr. Patterson makes the statement that contracts are going to reassert themselves. As I read S. 2387 (the vertical divestiture bill) that would be illegal. It specifically denies contractual or financial arrangements which could be viewed as a reassertion of vertical integration. Thus if contracts did reassert themselves, they will do so illegally under the terms of that bill.

Refinery financing
I would also like to say something about the refinery in Louisiana, which was built without a guaranteed access to crude oil. I know exactly how that refinery got its financing. It did so because a provision was put into the regulations which stated that a new refinery would receive an allocation under the allocation program. In other words, we may have adequate refining capacity available at the present time but if someone wants to build a new refinery, all the other refineries have to divvy up their crude oil to that new refinery. So, in effect, other refiners had to guarantee the supply of the Louisiana company.

James M. Patterson

The balance of integration
I'm not proposing that the majors be prevented from integrating foreign marketing or from integrating horizontally. I think it would be difficult to establish a case of horizontal monopolization in the industry — it just is not there, as all of the speakers this morning, and congressional testimony during the last three or four years have indicated. From a classic concentration or rational standpoint, the industry looks great. It looks better than most. The problem is not with vertical integration per se but rather with the balance of integration, in which an overwhelming share of the industry is integrated, and a very small part of crude end product moves into non-integrated channels. It is the pervasiveness of vertical integration rather than horizontal concentration that I think is the difference.

There are obviously many people who disagree with that.

I am not trying to restrict growth in any one area, nor would I want to restrict Mobil from going into merchandising. Capital ought to be free to move where it wants to go. But crude oil becomes a very special problem, when you have balanced vertical integration. I think as long as the industry is balanced, you are either going to have regulation or divestment. I do not think we can return to the heyday of the post World War II period; the public will not allow the industry to write its own future course as long as crude oil is integrated in a balanced way and therefore really not available to all buyers.

William A. Johnson

The public was all for throwing Christians to the lions in Rome. I don't think we should support what the public wants just because the public wants it.

James M. Patterson

You never ran for office.

6

THE INTERNATIONAL IMPLICATIONS

KANGAROOS AND WOLVES: DIVESTITURE AND OIL PRICES

Thomas R. Stauffer
Professor,
Harvard University

I have been asked today to discuss the international implications of divestiture, with particular reference to the effect of vertical "disintegration" of the oil industry upon OPEC and the future price of oil. It is tempting to describe this proposal as a "non-remedy to a non-problem" and quit. However, more precisely, I propose for the next ten minutes or so to deal with the implications of divestiture as a "non-remedy to a real problem."

Since the issue itself is miscast, this question is difficult to address. My task in discussing divestiture as a device to weaken OPEC is akin to the dilemma of the man who was once sitting in a bus. The passenger in the opposite seat periodically rolled up small pieces of paper, opened the bus window, and then threw them out. He did this every few minutes until the first chap, quite mystified, finally leaned over and asked: "Hey, buddy, why are you doing that?" The other replied, very condescendingly: "Well, you see, it keeps the kangaroos away!" The other passenger was puzzled by this comment: "But, look, there aren't any kangaroos around here." Whereupon the second fellow looked up, beamed from ear to ear and said: "Yeah, that's right!"

The proposal that vertical divestiture of the oil industry will lead to cheaper oil is a dangerous delusion, just like the use of rolled paper talismans to ward off kangaroos. It is dangerous because at best it is a non-remedy to a real problem, and it is even more dangerous because it most probably will exacerbate problems of price and supply stability leading to still higher prices.

Professor Chandler mentioned earlier today that history has some utility other than amusement — one such use is the hope that history should help prevent us from repeating mistakes. Here, in the context of the international implications of divestiture, some understanding of recent history might prevent us from making the kind of mis-diagnosis which could lead to a false prescription.

In particular, with regard to oil prices, it is vitally important to understand that the evolution of the OPEC cartel, and the dramatic increase in oil prices, were preeminently political phenomena, not economic events. Consequently, economic rearrangements, such as the proposed economic restructuring of the oil industry are quite immaterial within this political context. Hence, if we misunderstand the disease by misreading the symptoms, we may all too easily end up with a solution which is either useless or perhaps harmful.

Let us recapitulate the recent history of oil pricing. Two distinct phases are discernible, and both are the proximate consequences of political conflict in the Middle East. First, the closure of the Suez Canal in 1967, originally intended as part of Israel's economic war-fare campaign against Egypt, set the stage for Libya's price increases in 1970 and 1971. Short-haul crude oil provided the leverage, and opposition to "Zionist-imperialism" provided the will. Politically conditioned and taking advantage of a political development, the Libyans triggered the series of price increases which subsequently became auto-catalytic.

Similarly, the price increase in December, 1973, although initiated by the Shah of Iran, was a consequence of political events. The Arab embargo had reduced oil supply, and the shortage caused widely-publicized but no less spurious high prices in the spot markets for crude oil. These politically-induced price increases proved irrevers-ible, again because of lack of a political settlement in the broad sense under which the Saudis might have reversed the Iranians.

Finally, we must recognize that cheaper oil cannot result from any action by or upon the oil companies. It may come, if at all, only out of some kind of political compromise within the framework of a more general political settlement in the Middle East. Indeed, the relation-ship is symmetrical; political conflict precipitated higher prices, and political settlement might have lowered them. It is reported that the

Saudis during 1974 offered the U.S. a lower price for oil — some $4 per barrel or 10 cents per gallon — in return for our supporting U.N. Resolution 242.

It is my point, given the political antecedents of today's oil prices, that it is very, very difficult to construct any plausible scenario whereby divestiture of the international oil companies can affect the price of oil except possibly in the wrong direction upward — a point to which I shall return in a minute. Oil pricing today is decoupled both from the number of buyers in the market or their character; the presence, absence or dismemberment of the major companies is immaterial as far as any potential downward pressure on prices is concerned. They exercised surprisingly little market power during their heyday; today they enjoy still less. Today's price of oil is a political price, not an economic price; it is not the major companies which are important factors but the major powers.

One could argue the opposite, i.e. that the presence of smaller companies was destabilizing, offering more power to host governments. Indeed, to pursue that consideration further, we may look again at the history of oil prices, following Professor Chandler's model. It was the presence of the smaller companies, such as Occidental, which provided the necessary leverage for the Libyans in 1970 and 1971. If the Libyan government had faced only the major oil companies with their wide geographical diversification and financial balance, its ability to pull off the coup which then triggered the subsequent price increases would have been very much less. It may well have failed. A proliferation of buyers reinforces the bargaining stature of the cartel, rather than weakening it.

Further, once we recognize the political genesis of oil prices, we understand why OPEC's offer curve is flat. (Incidentally, I use "OPEC" here as a shorthand for Saudi Arabia, Kuwait, and Iraq — the only three countries which are in a position to increase supply significantly.) The level of this offer curve — the price — is set by politics, the Saudis' willingness to restrain prices being conditioned by the U.S.' policy towards Jerusalem, while the flatness of their curve is governed by their geological endowments and conditioned by "petrodollar" reserves. Nothing which we might do vis-a-vis restructuring the major oil companies will have any impact whatsoever on the nature of that offer curve.

"OPEC" — referring still to the three swing producers — had devised the ultimate prorationing scheme. This is their own version of the old "take or pay" agreement; now one has the option of "taking or else." If one does not like their terms, their posture is: *"malesh"* — no deal. Given this new configuration, divestiture is immaterial. Anyone who claims that the proliferation of buyers resulting from divestiture will affect OPEC's offer curve *downward* is deceiving himself and, what is worse, deceiving us all.

In fact if we pursue this argument further, we suspect that divestiture may really have the opposite effect. Whereas there is no advantage to be gained internationally, i.e. with respect to the global oil price level, it may well follow that there are two distinct disadvantages — again from the international standpoint — accruing from a policy of divestiture.

The first such scenario deals with the possibility that divestiture will in fact increase the strength of OPEC — i.e. the "big three" — and consequently increase the probability of still higher prices. The second focuses on the likelihood that divestiture might over the medium- and longer-run reduce U.S. access to those sources of foreign oil upon which our economy must necessarily depend for the next 10-15 years. Given that our increasing dependence upon imported oil is the unshakable legacy of the past, the question of access is crucial. Access, in turn, given recent tendencies among the other industrialized countries, may depend increasingly upon the flag of the company which produces the oil or operates the fields.

Let me try to spin out these two scenarios, treating both as hypotheses, which are to be opened for discussion after Clem Malin has finished. My crystal ball is no better than anybody else's. It's cracked, it's dirty, and it may even be upside down. But let's see if we can construct two hypothetical scenarios, one dealing with prices and one dealing with availability.

First, with regard to OPEC strength and divestiture, we note that there is indeed an interface here between domestic energy development and the international scene. If there transpires what we suspect after divestiture, and if there then would ensue a period of five or ten years when corporate expectations were blurred, blunted or perhaps totally eliminated, then one must fear that the requisite investments in domestic oil and gas would falter or cease. It is highly probable that as a consequence of divestiture that U.S. domestic energy production (outside of the nuclear field) will fall even more markedly.

Any further perverse regulatory responses to that would simply accentuate decline. Then, following our scenario to its next step, the confusion and chaos consequent upon divestiture is therefore likely to increase our demand for foreign oil quite significantly. Every single barrel demanded incrementally from OPEC at the present time strengthens its hand; the perverse dynamics of the system are such that increased imports during the confusion resulting from divestiture simply consolidate OPEC's position, which is indeed the opposite of what we have been told to think. That's the first scenario.

The second deals with the question of availability of oil and access to it, particularly as a consequence of the developments of the 1973 embargo. There is, and there will be, a correlation between the flag of the company producing the oil overseas — whether within OPEC countries or within future OPEC countries — and the security of access of that country of domicile to the oil. In 1973 the U.S. companies managed, through their global logistical structure to divert significant quantities of oil to the U.S., more than otherwise would have been our share under the terms of the embargo imposed by the Arab states. In retrospect this act of loyalty to the United States may have been a strategic error for the oil companies, since it was an act that found no resonance here.

The consequences of this act of diversion did not escape the governments in Western Europe, nor did the Japanese fail to notice what was going on. So, today, we see that our allies, the other OECD members, far from concerning themselves with dismembering or restructuring or "disintegrating" their oil companies overseas, are marshalling a variety of devices designed do exactly the opposite. Their goal is to strengthen their flag presence in any new area in the world where they think, or hope, that they can find oil. One of the stipulations intrinsic to these programs is that any oil which is found by a Japanese or French or other company must be dedicated to the home market; this is now often embodied in the agreements with the host countries.

We have set up our predicate — the fact that our international competitors and allies are systematically bolstering their own oil firms and abetting their ventures abroad. If our own firms are dismembered, stripped of the economics of logistical balance, while simultaneously facing tax disincentives, their ability to compete for new acreage is compromised. Among other effects, this would greatly reduce U.S. companies' ability to diversify geographically, a prime strategic consideration today. If long-term contractual relationships were also to be excluded — as surrogates for integration — the impact would be still worse. Thus, at a time of global concern for national access to overseas natural resources, divestiture weakens our own hand precisely while allies are doing the opposite.

On balance, therefore, the proposal that "disintegration" of the oil companies will somehow cut the price of gasoline is like the idea of getting rich by betting on a bad wheel at Las Vegas. At best, one might break even, but most likely one will lose heavily — and in a game where debts are enforced!

In closing let me return to our original zoological analogy — there seems to be an inadvertent zoological undercurrent in today's sessions. Divestiture in the interest of cheaper oil is like warding off kangaroos with rolled paper. First, there "ain't" any kangaroos out there. And second, while we are busy rolling paper, Jim Akin's wolf may well slip up and bite us again.

DIVESTITURE AND THE POWER OF OPEC

Clement B. Malin
Assistant Administrator
International Energy Affairs
Federal Energy Administration

It is a pleasure to be here today to discuss a very important, but largely ignored, or perhaps forgotten aspect of the current debate on divestiture, the international implications of breaking up the oil companies. I regret I have not been able to spend the whole day here, because I hope and assume that speakers have been addressing the many substantive questions about divestiture from the domestic perspective. The domestic ramifications of divestiture have been getting most of the public attention. So far, almost the only international justification one hears is that (and here I quote from the Report of the Majority Staff of the Senate Antitrust and Monopoly Subcommittee, Page 13): "Ultimately one of the most important effects of divestiture will be to place restraints on the pricing power of the OPEC cartel." I'd like to examine that hypothesis today, by raising a number of questions — and attempting to answer some of them — questions that must be at least considered before an intelligent decision can be made on divestiture:

    — What are the real international goals of divestiture legislation? Is it aimed at the OPEC cartel? If so, will it . . .
    — Increase international production?
    — Lower world oil price?
    — Increase supply security for the U.S. and other nation?
    — Lessen the power of OPEC?
    — If it is aimed at the companies, is the intent to:
    — Increase domestic production?
    — Increase international competition?
    — Lower prices to U.S. consumers?
    — No matter which target, what will be different if divestiture is implemented? Why?
    — In short, what is the situation now, what are our goals and the rationale for change, and what would be accomplished if such sweeping changes as are implied by divestiture were introduced?

Let's look at some of these questions.

The free world now uses about 50 million barrels of oil per day, some 60 percent of which is produced in OPEC nations. The international oil industry keeps 800 million barrels of oil moving at all times, and lifts, transports, refines, and markets nearly 80 percent or 9 billion barrels per year of OPEC oil for end-use consumption. The U.S. alone consumes about one-third of the world's daily oil production, 16 MMB/D. Our total oil imports (crude and product) amount to 6.5 MMB/D currently, virtually all from OPEC sources. Further, U.S. crude imports from Arab nations have gone from about 22 percent before the 1973-1974 embargo to about 45 percent now. Saudi Arabia has become the leading source of our crude oil imports. As a matter of fact, Saudi Arabia has been the Number 1 supplier of U.S. crude oil imports since November 1975 and was a close second for the whole of 1975, and probably would have been two in 1974 but for the October 1973 War and the embargo.

The recent, widely publicized figures showing imports exceeding domestic production may be considered a fluke but they are indicative of an overall trend. The trend should be of concern. And it looks as if it is not going to be reversed quickly.

The President has proposed a program of conservation and resource development, which, if implemented, could give this country sufficient energy security initially by 1980 and more by 1985 to sustain interruption with minimum economic disruption. Unless major portions of this program are adopted, however, FEA projections of future demand or OPEC oil indicate that the United States could be importing more than 7 MMB/D in 1980 and as much as 9 MMB/D in 1985. It is also projected that as much as 4.1 MMB/D (or 55 percent of projected total U.S. import demand) will come from Arab sources in 1980 and as much as 5.5 MMB/D by 1985. U.S. strategic reserves are scheduled to total approximately 325 million barrels by 1980 and 500 million barrels by 1985. At the projected import rates, the reserves would cover an import interruption of less than 2 MMB/D for 6 months in 1980, and about 5 MMB/D for 3 months for an import interruption in 1985.

I mention these points because it is important to appreciate the significance of the extent to which the U.S. may have to depend on international supplies of oil in the future — if we are unwilling to commit this nation to a significant reduction in our level of imports. Under such circumstances, the role of the International Oil Companies (IOCs) may be crucial to the security of our import supplies and those of our allies.

Until recently, the world oil market was dominated by International Oil Companies (IOCs) headquartered in the U.S. and one or two other major consumer countries. Those conditions generally assured a secure supply of oil at predictable, low and stable prices, because under the concessions the IOCs determined the rate of development and production as well as the price of crude marketed internationally. Moreover, the existence of and the IOC's assured access to, and control over excess production capacity in various oil producing countries, provided the supply security for the adequate, stable, uninterrupted volumes of petroleum so vital to the economic development of both the oil producers and consumers of the free world.

That control, together with the fact that the United States had an export capability, rendered the oil supply disruptions of the 1950's and 1960's ineffective and short-lived.

But some of these conditions have changed very significantly: The United States is now a net oil importer; and while there is a substantial amount of excess production capacity in oil exporting countries, it is no longer under the control of the IOC's. Moreover, the principal reason such excess production capacity exists is because the oil exporting countries have shut in production to sustain a world price more than five times the 1973 level.

The price, the terms of access and the production levels for the international oil market are set by the OPEC member states. Their own national oil companies are moving to establish refineries and related facilities in their own countries to market petroleum products internationally. And a few have sought to invest in such operations, via joint ventures, in consuming countries. These changes, as well as the relative dearth of alternative sources of supply mean that we can probably expect continued upward price pressure from OPEC and possibly even some production cutbacks — whether deliberate and selective or unavoidable and general.

But these changes notwithstanding, the IOC's are still important to the commercial marketing of OPEC oil — and to that extent they continue to exercise some influence in that market. The embargo and production cutbacks of 1973-1974 demonstrated the extent of OPEC control over world oil prices and over-supply to the entire system; but the impact of that cutback in supply demonstrated the inability of OPEC to control whether or not a specific nation received oil. This is an important weakness in the capability of OPEC to selectively target production cutbacks on particular countries.

The role of the IOC's is crucial to an understanding of the reason for that weakness; and their continued control and management of the international distribution and logistics system, as well as their equity interests in the refining and marketing of international oil, are the principal components of the IOC role.

Now, having set the overall context, let's focus on some of the questions and implications raised by proposals for divestiture of international oil industry operations. We'll concentrate primarily on vertical divestiture, because horizontal divestiture's international impact seems to be less direct.

First, there are legal and legislative questions about the expectation of enforcement of divestiture of the international operations of the oil industry.

Proponents of vertical divestiture may intend to require its world-wide application, but the legislation as it stands now contains no hint as to how this is to be accomplished. Simple application of the domestic divestiture proposal internationally would multiply the number of U.S. companies, though it might not radically increase the number of U.S. buyers confronting the same small number of OPEC sellers. At the same time, the commercial competitive capabilities of those U.S. buyers relative to those of other private and national integrated oil companies around the world could be reduced. Potential purchasers (whether in larger or smaller numbers) can enjoy lower OPEC prices only if OPEC nations compete with each other, or with other alternative suppliers, to sell more and more for less and less. Control over supply is the key, rather than the number of bidders in the market. OPEC members have not shown a great willingness to cut prices to compete with each other. If alternative non-OPEC sources of supply could be developed, then OPEC nations would face some greater degree of selling competition. But, if divestiture served to introduce additional uncertainty into the flow of U.S. company investments in exploration and production in the developing countries, then the Arab producers of OPEC might remain the only possessors of "excess capacity," maintaining their control over supply and price. Would development of alternative additional sources slow down under divestiture? It's difficult to know in advance. If it would, however, what impact would it have on our relationship to OPEC? Would the United States benefit? Would our Allies?

Further, within the major foreign consuming areas, U.S. divestiture would probably give impetus to development of national vertically integrated firms. In Europe and Japan vertical integration is seen as a means to security and efficiency. If U.S. firms cause them concern it is because of the implications of their political and economic dependence. Divestiture would probably be seen as an opportunity, if not a requirement, to promote national vertically integrated alternatives.

There is another possibility with respect to international divestiture, what if the legislation were to allow or require the affected companies to separate their U.S. components from their international components, and so limit the impact of legislation on their company to the physical assets located in this country? Would this happen? "Shore-line" divestiture has been suggested by one proponent. Therefore, we should think about why the companies might attempt "pre-emptive divestiture," and what would be the possible outcome.

Each of the 18 companies "targeted" in the proposed legislation would have to make its own calculations of the desirability of such pre-emptive divestiture, but if a company could, and if the economics showed it should, then it probably would follow that course. The result could be a new set of integrated international oil companies based outside the United States; a number of U.S. production companies, pipeline companies and refining and marketing companies unconnected to each other or to the few non-U.S. integrated internationals; and, in some cases, a new non-U.S. international with one U.S. component (probably either production or refining-marketing, depending upon the company's perception of future profitability).

To sketch the case quickly, of the 7 largest oil companies in the world:

— 5 are U.S. based, but
— only 15 percent (3.5 MMB/D) of their world total "controlled" production (22.6 MMB/D) is in the United States
— only 23 percent (6 MMB/D) of their world total refining (25 MMB/D) is in the United States
— all depend heavily on OPEC crude even to supply the U.S. market
— only a small part of the world tanker fleet is officially U.S. flag or U.S. owned, but a large part of the total fleet is effectively controlled (owned or long-term leased) by U.S. companies.

Thus, if it were not precluded legally and if it were worthwhile commercially, the companies could protect the largest part of their assets through pre-emptive divestiture.

This is not to say that the companies targeted by the legislation would escape unscathed competitively through self-imposed divestiture. Even the majors would be weakened relative to other "foreign international oil companies." It is worth asking whether or not U.S. interests would be served as the companies were weakened.

Where would the companies move? Canada? Britain? Norway?

Japan? The Bahamas? Iran? Who is to say? The point is that those companies would have international markets in many other parts of the world, and would almost certainly have to rely heavily upon OPEC for production. Could they be persuaded to concentrate new resource development there? That would depend upon where the profits and long-term outlook would be best. But surely the decisions of those "formerly American" companies would not be overly circumscribed by a feeling of great indebtedness to their former host country.

I have already noted the reasonable equity and efficiency with which the IOC's allocated available oil during the 1973-1974 embargo; despite the desires of France, for example, which wanted special treatment from both the Arabs and the oil companies. The OPEC nations had not targeted France, but the companies provided only a proportionate share of the reduced supply. Thus, the targeted countries (especially the U.S. and the Netherlands) did not suffer disproportionately, nor did other nations benefit unduly.

But since the last embargo, an International Energy Agency has been established and has put into place an International Emergency Program (IEP). The IEP depends upon the ability of the oil companies to manage supply and distribution, presently the main lines of defense in the event of another embargo or supply disruption. Is it the case that in future emergencies the limited supplies could still be directed as easily among separate companies divested on functional lines as within single integrated firms? Would the IEP Phase I (voluntary allocations by and within each company network) be effective or would consumer nations be forced to rely on Phase II, or even perhaps the rationing and stringent mandatory actions called for under Phase III? Would U.S. divestiture, by reducing the ability of the oil industry to cushion the impact of future embargoes, impact directly on the security and political/economic interest of our IEA partners? These are questions that should be of concern to us and to our fellow members of the IEA.

But beyond that, would the ability of producers to target and embargo be reduced or enhanced by U.S. divestiture legislation? Would divested U.S. domestic companies, and the non-U.S. internationals with whom they deal through sales be easier to police? Would that ability, almost totally lacking in 1973, be much more real if there were non-U.S. based companies with worldwide networks, a much smaller part of the world tanker fleet under effective U.S. control, and no international oil companies with downstream market interests to protect in this country? Under the best conditions, at the very least the new international companies would have to be brought into the IEA — that is if they remained based within IEA countries rather than moving to OPEC nations.

Further, and on a different scale, it should be remembered that the new U.S. refining-marketing companies, though forced by divestiture to find new supply sources, would still seek assured access to supply. State oil companies of the producing countries and the integrated foreign oil companies could probably fulfill this role without being in illegal "control" of these refiners. The U.S. production/service companies, operating on small margins and perhaps burdened with corporate debt might not survive the transition. Producer governments might purchase them outright or invite foreign integrated firms to take over the technical/marketing role.

At the same time, the U.S. law could have resulted in disruptions, forced breach of contracts (despite foreign law covering the same contractual relationships), and the possible impairment of this country's ability to compete for and develop new sources of energy and of revenue, including exploration and production in the developing countries and the present "non-oil LDC's." Already regarded as high-risk ventures, investor interest in such activity is to expand world production and diversify sources of crude, thus enhancing supply security. If a refinery's — or a nation's — supply security is not enhanced, it is generally diminished. It rarely remains static. This potential outcome must be considered.

Divestiture would render questionable the utility of many of the arrangements now evolving between major producing countries and U.S. international oil companies. The production/service companies in the OPEC countries might not be able to fulfill the envisioned long-term supply offtake role through divested refining marketing segments. At first blush, this might not seem to matter to nations. But, it should: Aramco, for instance, will be involved in between 20 and 25 percent of all OPEC oil exports. Saudi Arabia will have most of the excess production capacity in the world. The mere size of the reserves, the excess capacity, the size of the U.S. market, and the potential and proven ability of international integrated companies to allocate around "targeted" countries in an embargo make this a supply security consideration that could deteriorate if the companies were broken up. At the present stage of the negotiations, the owner-companies of Aramco are trying to get a supply commitment from Saudi Arabia. If they are talking about 7 MMB/D or so, and if those companies are then cut off from or given sharply reduced access to U.S. markets, what will be the impact for U.S. supply security? Even if the Aramco partners don't buy so much, other companies may, of course — but will the price to them be higher or lower? What would a changed trade flow mean to our allies?

None of these questions can be answered with total certainty, and none of the potential outcomes can be predicted absolutely, but all must be weighed before an intelligent decision can be reached on the overall issue of divestiture.

On these points, some might argue that the price is worth paying — that forcing arms-length dealings between U.S. refiners and OPEC may be an advantage. This is a critical judgment. It assumes that 18 or 22 or some other number of refining companies without involvement in producing countries (although some might be linked with producing countries refineries) somehow maintain or increase the degree of national and international supply security and, at the same time, can drive a harder bargain on price with those governments than the presently integrated firms. But real supply security rests on excess capacity and market access, not "hard bargaining" — OPEC has most of the world's excess capacity and they appear to have the will to keep it shut in to sustain prices. If additional supplies are not brought on in countries able and willing to maximize production — whether in the U.S. or not — then OPEC's power to unilaterally set prices will not diminish. We need competition among oil producers, not among consumers facing a strong producers' cartel. And the availability of those alternative supplies will depend in large part upon the role of the IOC's in exploring for and producing oil from such sources; and their incentive to do so will depend importantly on assured access to markets.

With or without formal agreement successful prorationing of OPEC production rests upon the continued willingness of Arabian peninsual producers — especially Saudi Arabia — to hold back production disproportionately for the sake of the cartel. The smaller the market for OPEC oil — whether due to reduced demand or additional supplies from alternate sources — the more costly and difficult becomes a prorationing scheme. Would divestiture indirectly or directly alter total demand for, or total supply or, OPEC oil? I think not. The key question for the swing producers would remain unchanged after divestiture as before. What is the economic and political value of the cartel to them in maximizing the advantage of their single resource? That value seems to be high.

The key question for the United States is: What would happen to our efforts to reduce our dependence on imported energy or even efforts to diversify sources of supply? Through divestiture would we be advertising the fact that we did not consider them worthwhile from a national policy perspective? This is a crucial question.

There are a number of other questions as well — for example, what would be the reaction of the British Government if the British Petroleum share of Alaskan North Slope oil were placed in jeopardy? Again, I don't know. But I know that we should think about it.

My intent today has been to suggest some of the many and complex questions that must be asked and answered about the international implications of divestiture. We can not consider divestiture a purely domestic political issue. We must identify and evaluate the international costs and benefits of divestiture and then fit them together with those costs and benefits strictly associated with domestic concerns. We must consider the effects: on the U.S. and the IEA countries and our relationship with OPEC; on national and international supply security; and upon the price of oil in the world and in this country. And all of these questions must be answered in the context of the world supply/demand outlook for the next decade or beyond.

There are great numbers of important issues yet to be addressed. Perhaps we have made a start on that today. Thank you.

7 DIVESTITURE AND PUBLIC POLICY

John C. Sawhill
President
New York University

I am delighted to have an opportunity to meet with so many energy experts today, but I should begin by expressing my biases. Frankly, I am disappointed that so many knowledgeable people have spent so much time and effort discussing divestiture. Actually, I feel that the quality of the discussion has been good, and some very cogent arguments have been advanced by almost every speaker. But, the thing that troubles me is that we are debating an issue which is not very important for America's energy future. I wish we were devoting the same effort to some other more important issues; and, in particular, I wish we were spending the day discussing an issue which will assume critical importance in the next few years, namely, the question of how to conserve on energy resources more effectively.

But, to return to the divestiture issue and summarize briefly. On the question of vertical integration, the facts presented by those for and against divestiture are surprisingly similar. The argument seems to turn on the costs and benefits. I am not sure that the costs are quite as severe as Annon Card pointed out. For example, I am not really sure that we would lose a million jobs or that GNP would drop quite as sharply as he indicated; but, on balance, when you weigh the management time which would be consumed in divestiture proceedings, the legal battles which would ensue, the disruption and dislocation which would occur, the uncertainty that Bill Johnson described, the financial problems that Gerry Parsky described so well, I think the arguments against divestiture are persuasive. One interesting point which came out in the discussions was the international implications of divestiture and the impact on OPEC that divestiture might have. Based on this discussion, I would conclude that there is probably more evidence that divestiture would strengthen the cartel rather than weaken it.

Mr. Measday in his presentation argued that ten or twenty years from now prices would be lower if the industry was forced to spin off downstream operations. And I suppose that you could go through a line of theoretical economic reasoning to support that statement. But, since it is very difficult to make a good case that profits in the oil industry are out of line, I am not convinced. As a matter of fact, I pulled some statistics together which show that the largest U.S. oil company in terms of gross revenue, Exxon, actually ranked 161st among U.S. corporations in terms of profitability. The second, third, fourth, and fifth largest oil companies in terms of revenue ranked 252nd, 327th, 352nd, and 455th. So, it seems to me that it is difficult to argue that prices (and hence profits) are inflated.

A second argument which was advanced hinged on the ease of entry or lack thereof. Certainly, this is one of the criteria by which the structure of the market should be evaluated. But, I question the conclusion that entry is overly restrictive. It is interesting to note that in the discussions of horizontal integration, the argument about the ease of entry took on a different character. It was suggested by those who are opposed to horizontal integration that one reason for doing so was to restrict entry into non-oil energy industries. Yet, restrictions of entry is the very thing that those who are concerned about vertical integration want to prevent. It is somewhat inconsistent to say that there should be easy entry in the oil industry but not in other energy sectors.

My conclusion, then, is that at a time when the U.S. needs to expand its energy supplies and improve the efficiency with which it uses these supplies, we should forget about divestiture. If we really have concerns about the ability of our antitrust laws to work effectively, or if we conclude that the FTC and the Department of Justice are not doing a good job, then Congress should turn its attention to strengthening our laws or the agencies charged with executing them. I am not in a position to know whether or not they are as effective as they should be. Mr. Anton and Mr. Parsky both indicated, however, that if antitrust problems exist, it is probably better to solve them through existing institutions, and I would have to agree with that position.

The discussion on horizontal integration differed somewhat from the discussion on vertical integration in that there was some disagreement on the facts, at least the facts on total U.S. energy reserves. Mr. Slick said that the oil companies owned 19 percent of U.S. coal reserves, and Senator Haskell said the figure was about 57 percent. Obviously, that difference is quite large, but I am not sure it is terribly relevant because when you get through with all of the discussion on the concentration ratios, it seems to me that we are really talking about a question of political philosophy; a question, if you will, about what role we see the government playing vis-a-vis private industry in this country.

I thought the discussion of whether in fact they would have interfuel price competition or not was interesting, but if you look at other industries where more than one competing product is sold by the same company, I think you generally find evidence of a high degree of competition. Certainly, the heads of Chevrolet and Pontiac divisions in General Motors compete effectively against each other. So, I tend to discount the argument that interfuel competition will be reduced if horizontal integration is not prohibited. I would much rather see the oil companies, with their technological and managerial capabilities, developing America's energy resources than developing America's retail or real estate resources. Perhaps, if you project the trends out twenty or thirty years in the future, which is a hazardous occupation at best, you could make a theoretical case that the oil industry would dominate all forms of energy. But, with equal logic, you could make the case that IBM or any other fast-growing company is going to gobble up its industry. The facts suggest, however, that oil industry domination of other energy sources is not causing serious problems today. If it becomes a serious problem, we probably have adequate institutions in government to deal with it. In the meantime, we have important tasks to do in this country in terms of developing our energy resources. I think we ought to get on with the job of doing so.

As I said at the outset, however, it seems to me disappointing that we are sitting here debating this issue, because it means that we are not debating the issues which I feel are more relevant to America's energy policy. Several of the speakers have addressed the question of why this issue has come to the fore. I think the answer has been properly given that Americans have lost confidence in their institutions, whether they be business, government, or even universities. As a matter of fact, a recent Harris poll indicated that ten years ago 55 percent of the American people said they had a great deal of confidence in major business organizations. Today that figure has dropped to 16 percent. The comparable figures for Congress are 42 percent to 9 percent. I was on a panel with Lou Harris the other day, and he reminded me that university presidents are now down to 31 percent from 70 percent. So none of us are immune to the crisis of confidence, and it is going to take a major effort to restore that confidence.

Congress has latched onto the divestiture issue as a way of diverting themselves from dealing with the more fundamental energy issues, namely, the problem of what we must do to expand energy supplies and what we can do to curtail energy demand. It is clear to me that our energy policy must encompass *both*. Some have suggested that very strenuous energy conservation measures would solve our energy problem. But, careful analysis indicates that such a policy, as desirable as it is, will not do the whole job. The recent FEA national energy survey (and most people I have talked to generally agree with the figures in the report) shows that if the energy prices continue at current levels —that is, if they continue to move with inflation — energy demand will probably grow at about 2.8 percent a year. Of course, the growth rate will vary by fuel, but the 2.8 percent figure seems to be generally accepted for overall demand growth. With some fairly strong conservation measures, we could probably reduce that 2.8 percent demand growth to 2 percent, or maybe 1.8 percent, or maybe 2.1 percent — somewhere in that range. But, even a 2 percent growth rate would still leave a fairly wide gap between supply and demand — so it is clear that we have got to take measures to expand supply if we want to reduce imports — or even hold them even.

As we all know, there are a lot of problems associated with expanding supply. Oil and natural gas supplies are running out — so naturally, additional drilling is only a short-term solution. Coal has a lot of environmental problems associated with expanding production, which suggests that it will be difficult to accomplish. I will not dwell on the difficulties associated with expanding our nuclear capacity. Clearly, the estimates of growth in nuclear generating capacity now being made are significantly lower than they were a few years ago. Finally, there are the "exotic" sources, but as we all know, they will not solve the near term problems.

For the immediate future, Congress will have to take steps to permit prices to increase so that there will be appropriate incentives to expand supply. One such step is the deregulation of new natural gas prices at the wellhead. Another is passing legislation which, instead of rolling back energy prices, would permit them to rise to world levels. Congress will also have to take steps to remove some of the institutional barriers to energy conservation.

Earlier in the day, Gerry Parsky argued that we should not "artificially" push prices up. I am somewhat puzzled by that statement. If we assume that artificial means something other than what would occur in a free market, it is clear that there are already a number of factors at work which cause prices to move up or down other than supply and demand. In some cases — such as tax incentives and disincentives — the objective is to reflect social costs or benefits in the price structure. Thus, we might conclude that it would be good public policy to reduce congestion and pollution by reducing the use of gasoline. One way to do this would be to raise the price of gasoline via the excise tax to reflect the cost of congestion and pollution. At the present time, we have an average 12-cent excise tax on gasoline when you combine federal, state, and local taxes. Each cent increase on the gasoline tax puts another billion dollars in the federal treasury. Thus, this tax is a good revenue producer, and the undesirable impact of an increase on lower income groups could be offset by refunding the increase to them in a variety of ways that would not be administratively complex. The tax could be increased over a period of time, if policymakers feel that a large, one-time increase would be too disruptive. And, the revenues from such a tax could be used to improve substantially public transportation.

Most of the studies I have seen indicate that it is very difficult to get people to use their automobiles less if you only improve public transportation. There has to be a disincentive to use automobiles, as well as incentives to use some other form of transportation, in order to really make much of a difference. I would suggest, therefore, that Congress reorder its priorities and put a gasoline tax increase at the top of the agenda, rather than divestiture. I would submit that such a tax, combined with the other measures I mentioned earlier which would permit prices to rise above the levels now imposed by FPC and FEA regulation, would be a much more constructive way to deal with the energy problem. Of course, there are a variety of other conservation measures which should also be enacted — i.e., providing tax credits to people who insulate their homes, providing tax incentives for recycling and reuse of materials, etc. Just expanding the Oregon ban on nonreturnable bottles nation-wide would help. And, certain steps must be taken to permit energy development to proceed.